Polyhedron Models

Polyhedron Models

Magnus J. Wenninger

Cambridge   at the University Press   1971

Published by the Syndics of the Cambridge University Press
Bentley House, 200 Euston Road, London N.W. 1
American Branch: 32 East 57th Street, New York, N.Y. 10022

© Cambridge University Press 1970

Library of Congress Catalogue Card Number: 69–10200
Standard Book Number: 521 06917 3

Set at the University Printing House, Cambridge
Printed in the United States of America

In grateful memory of T. G. who tutored me in
the Philosophy of Mathematics

Mathematics possesses not only truth but supreme beauty, a beauty cold and austere, like that of sculpture, sublimely pure and capable of a stern perfection, such as only the greatest art can show.

Bertrand Russell

Contents

Frontispiece

The three regular solids whose faces are equilateral triangles—the tetrahedron, the octahedron, and the icosahedron—are paired with the three polyhedra derived from their symmetries.

If these regular solids are enclosed in a sphere and their faces projected on to the surface of the sphere, they generate a network of spherical triangles. These spherical triangles can be subdivided by the intersections of the planes of symmetry of the solids with the surface of the sphere. If plane triangles whose vertices coincide with those of the spherical triangles are allowed to replace them, the analogous network of triangles becomes the set of three polyhedra depicted here in colour.

These three polyhedra are representative or symbolic of the entire set of uniform polyhedra and the stellated forms presented in this book.

Preface

This book presents a well-defined set of geo-metrical solids, the seventy-five (known) uniform polyhedra, together with a representative set of stellated forms. A description of the underlying theory of polyhedra is included to bring out the relationships that exist between the various solids. But mainly this book is simply a set of instructions on how to make models of these solids.

The sources in which you can find an account of the mathematical theory of this topic are given at the end of the book. If in the past you found the study of geometry a bit difficult, or if at present you are not particularly attracted by geometry, you may wonder if this topic will hold your interest. The fact is that you really do not need to understand all the theoretical mathe-matics involved in the original discovery and classification of these solids. On the other hand you cannot avoid all the mathematics, especially the terminology used here and some of the sym-bolism.

The objective in this book will be to set down an explanation of the solids, at once simple and practical and not too speculative, one sufficient for the purposes of constructing the models. It is really surprising how much enlightenment will come, following the construction of the models rather than preceding it, and once you begin making them you may find that your enthusiasm will grow. You will soon see that each of these solids has a beauty of form that appeals to the eye in much the same way that the abstract mathematics appeals to the mind of a mathe-matician:

You may find the number of models presented here overwhelming, some of them extremely complex. Why should anyone want to make them? Maybe the answer is to be found in the reply of a mountain climber when he was asked, 'Why do you want to climb the Matterhorn?' 'The mountain is there, isn't it?' There is another question many people ask when they see these polyhedron models: 'What do you use

them for?' Maybe the answer to this is best given by a return question: 'Does beauty need to have uses?' Admittedly the only use a model has, once it has been constructed, is for display pur-poses. You can make some very attractive mobile models, and generally the constructions make lovely mantelpieces or centrepieces for tables at a banquet on special occasions. Stars seem to go with Christmas and here you have many star forms to choose from.

But on a more technical level you may have seen polyhedron forms used for space satellites. Then again the geodesic dome is found in archi-tecture and in engineering projects. Perhaps the polyhedral forms presented in this book have never been used simply because they have never been widely known.

I have myself constructed all the models pre-sented here and shown in the photographs. How long did it take me? My interest in this topic began in 1958 with a summer course at Columbia Teachers College in New York. During the fol-lowing year I made my first set of models, those given in section I of this book. My main source was *Mathematical recreations and essays* by Coxeter and Ball. Then between 1959 and 1961 I made all those in *Mathematical models* by Cundy and Rollett. Next I tackled *The fifty-nine icosahedra* by Coxeter, Du Val, Flather, and Petrie. I suc-ceeded in working out my own nets for each of these. The set graced the back wall of my mathe-matics classroom, growing as I completed each one between 1961 and 1963. The average work-ing time spent on each was about eight hours, plus three or four hours each to discover suitable nets. On the completion of this project I wrote to Professor Coxeter asking about *Uniform poly-hedra*. He kindly sent me a complimentary copy, one of three he still had in his possession. This monograph is a detailed account of the mathe-matical theory of uniform polyhedra. But for the purposes of making the models I inspected the drawings, done by J. C. P. Miller and col-lected together at the end of the monograph, to discover the facial planes from which I derived the parts. These facial planes are now being presented in this book. A set of photographs was also given in the monograph; these show wire models made by M. S. Longuet-Higgins, but they sometimes represent more than one polyhedron, so they are not the same thing at all as the models presented here.

My working time on the non-convex uniform polyhedron models varied greatly. The simpler ones took three or four hours each, the average would be near eight or ten hours each, a few of the complex models took twenty or thirty hours work. Two of the non-convex snubs required more than one hundred hours work each. Now that the work is complete, I must admit I myself am amazed. But the Chinese proverb applies: If you want to make a journey of a thousand miles, you begin by taking the first step. One step leads to the next, and soon the beauty of the countryside makes you forget the toil of the road.

A special word of thanks is extended to Mr R. Buckley for his truly remarkable calculations on the snub polyhedra and for his astoundingly detailed drawings of their facial planes. Without his help the book could never have been completed. Also a word of thanks to Dr H. Martyn Cundy for his deep interest in the book at all stages of its preparation, and to H. S. M. Coxeter, J. C. P. Miller and M. Longuet-Higgins, who did the original research for *Uniform polyhedra*, and who have provided the source of inspiration from which this book springs. Each in turn provided further encouragement and help to complete the task. Thanks are also due to Stanley Toogood for the photography, to the Syndics of Cambridge University Press for accepting the book for publication, and to the editorial staff of the Press who in an admirable way met the challenge of producing it.

Foreword

Interest in polyhedra runs through the whole gamut of intellectual activity from the two-year-old child who plays with wooden cubes to the mature mathematician who studies the subtleties of Branko Grünbaum's *Convex polytopes* (Wiley, New York, 1967). Some of the regular and semi-regular solids occur in nature as crystals, others as viruses (revealed by the electron microscope). Bees made hexagonal honeycombs long before man existed, and in human history the making of flat-faced solids (such as pyramids) is as ancient as any other kind of sculpture. The five regular solids were studied by Theætetus, Plato, Euclid, Hypsicles, and Pappus.

A considerable portion of the present book is devoted to 'uniform' polyhedra, which have the same arrangement of regular polygons at every corner. (Such a polyhedron is 'regular' if the polygons are all alike.) In any convex solid, a theorem of Euclid tells us that the angles at a corner must add up to less than 360°. After making a few models for himself, the reader will soon discover that the amount by which the angle-sum falls short of 360° is quite considerable when there are few corners (e.g. 90° for the cube, which has eight corners) but much smaller when there are many (e.g. 12° for the snub dodecahedron, which has sixty corners). This observation was fashioned into a theorem by René Descartes (1596–1650), who proved that the angular defect, added up for all the corners, always makes a total of 720°.

At about the same time, Johann Kepler (1571–1630) wrote an essay on *The six-cornered snowflake* (English edition, Oxford, 1966), in which he revealed his fondness for these figures by remarking (p. 37): 'Now among the regular solids, the first, the firstborn and father of all the rest, is the cube, and his wife, so to speak, is the octahedron, which has as many corners as the cube has faces.' It was Kepler who first published a complete list of the thirteen Archimedean solids, giving them the names by which they are still known. (The work of Archimedes himself had been lost, presumably in the great fire of Alexandria, which was so poignantly dramatized by Bernard Shaw in *Caesar and Cleopatra*.) Kepler also proposed the problem of enumerating the *isozonohedra* (convex polyhedra whose faces are congruent rhombi), and partially solved it by discovering the (first) rhombic dodecahedron and the triacontahedron. But his most important contribution to the ideas of the present book was his proposal to consider nonconvex polyhedra whose faces are stellated polygons such as the pentagram (fig. 21). He was probably unaware of the earlier work on stellated polygons by Thomas Bradwardine (1290–1349), who became Archbishop of Canterbury for the last month of his life.

Salisbury Cathedral is such a magnificent building, full of interesting relics, that many visitors fail to notice the tomb of Thomas Gorges, who died in 1610. The stone-carved decorations include a dodecahedron, three icosahedra, and two cuboctahedra, all 'skeletal' in the style of Leonardo da Vinci (1452–1519) who had made skeletal models of many uniform polyhedra using rods to represent the edges. A few miles to the south-west is the pleasant village of Wimbourne St Giles, where Antony Ashley was buried in 1627. His tomb is embellished with a truncated icosahedron, not skeletal but a solid looking just like the author's model **9**.

Since the time of Descartes, many other great mathematicians have contributed to this subject. Euler discovered and proved the famous formula

$$V - E + F = 2$$

which connects the numbers of vertices, edges, and faces of any convex polyhedron. Gauss used an irregular spherical pentagram (his *pentagramma mirificum*) to explain Napier's rules in spherical trigonometry. Cauchy proved that every convex polyhedron with rigid faces and hinged edges is rigid. Hamilton invented the Icosian Game (W. W. Rouse Ball, *Mathematical recreations and essays*, Macmillan, 1967, p. 262).

Von Staudt gave a new proof for Euler's formula. Schläfli extended the theory to n dimensions. Klein wrote a highly influential book called *Lectures on the icosahedron.* Fedorov returned to Kepler's problem of isozonohedra, discovering a strangely oblate-looking rhombic icosahedron; and Bilinski (as recently as 1960) completed the enumeration by discovering a second rhombic dodecahedron which would fit snugly into a box of unit height, breadth τ and length τ^2, where τ is the number $(\sqrt{5}+1)/2$ which belongs to the celebrated 'divine proportion' or 'golden section'.

In his infectiously enthusiastic style, the author gives clear instructions for making models of many kinds of polyhedra. These instructions are illustrated by photographs of his own collection, including what is almost certainly the only complete set ever made of the known uniform polyhedra. But photographs cannot really show the models in their full splendour. The most complicated 'snub' solids are not only extremely difficult to make but also highly decorative: a perfect instance of the connection between truth and beauty.

H.S.M.C.

Introduction: uniform polyhedra

If you are being introduced to this topic for the first time, your first question might well be 'What is a polyhedron?' You may recall that geometry itself is sometimes (not too exactly) defined as the study of space or of figures in space—two dimensional for plane geometry and three for solid geometry. The idea of sets is perhaps familiar also. If you use the language of sets, a plane figure may be defined as a set of line segments enclosing a portion of two-dimensional space. Such a plane figure is called a polygon. A polyhedron is then defined as a set of plane figures enclosing a portion of three-dimensional space.

All the terms used in this subject are derived from classical Greek. Plato, the famous Greek philosopher, left the imprint of this thought deeply fixed in Euclid's *Elements*. This ancient book, for centuries the only textbook of geometry, was concerned with 'ideal' lines and 'ideal' figures. The ideal lines are straight and the ideal polygons are regular, that is, they have all sides and all angles equal. The simplest regular polygon is the equilateral triangle. It is the simplest because it has the least number of line segments possible to enclose a portion of two-dimensional space. It is an interesting fact that Euclid's *Elements* begins with a proposition describing how to construct an equilateral triangle and ends with a study of the five regular solids. Each of these has regular polygons of the same kind for all its faces. They are known today as the five Platonic solids. The tetrahedron, which has four equilateral triangles for its faces, is the three-dimensional analogue of the two-dimensional equilateral triangle. It is the simplest polyhedron, since it has the least number of faces possible to enclose a portion of three-dimensional space.

With the equilateral triangle the following polygons enter the picture: the square (four sides), the pentagon (five sides), the hexagon (six sides), the octagon (eight sides) and the decagon (ten sides), all of course only as regular polygons.

Once you begin to make the models described in this book, you will quickly learn to draw all these figures accurately and will become acquainted with important properties belonging to each, especially the number of degrees in each interior angle. Not all the regular polygons are to be found in the regular solids; in fact only three are used. The hexahedron (six faces), commonly called the cube, has squares; the octahedron (eight faces) again has equilateral triangles; the dodecahedron (twelve faces) has all pentagons; and finally the icosahedron (twenty faces) has twenty equilateral triangles. Euclid's *Elements* closes with a proof that there are only five regular polyhedra.

A little experimenting with cardboard figures will soon lead you to see the reasoning behind a formal proof. Just as in a polygon two sides meet at a point called a vertex of the figure, so in a polyhedron two faces meet at or on a line (or in a line—the mode of expression is variable). Thus each face shares one of its sides as a line in common with another face. These lines are called the edges of the polyhedron. So each edge of a polyhedron belongs to exactly two faces and no more. The edges all meet at a point called a vertex of the polyhedron.

In the tetrahedron three edges meet at each vertex, or to put it another way, each vertex is surrounded by three triangles. It is enlightening to lay out these three triangles flat and to notice the sum total of the number of degrees in the angles at a common vertex. Three sixties give 180 degrees. If a fourth triangle is introduced the total is 240 degrees, but now you have a vertex of the octahedron. Introducing a fifth triangle gives 300 degrees, and you have a vertex of the icosahedron. A sixth triangle gives 360 degrees and you can see immediately that no polyhedral vertex arises. Everything stays flat.

Next you can try squares. A minimum of three is required, three 90s give 270 degrees, and a vertex of the cube can be formed. Adding a fourth square brings the total to 360 degrees and again you are left—flat. With pentagons the minimum of three will give you a vertex of the dodecahedron; four are too many, the total going beyond 360 degrees. With hexagons the minimum of three is already too many, three times 120 degrees. So no regular polyhedron exists with only hexagons for faces. And similarly for polygons with any greater number of sides. In this way you can see that only five regular solids are possible.

There is another set of solids known as the Archimedean or semi-regular solids. These all have regular polygons as faces and all vertices equal but admit a variety of such polygons in one solid. There are thirteen such solids and they are ascribed to Archimedes because he first enumerated them, although his work on them has been lost. References to his work on this subject are found in the writings of Pappus, a mathematician of the third century A.D. Kepler was the first of the moderns to formulate a complete theory concerning them.

The Archimedeans can be broken down into various subsets. There are first of all the five derived by the process of truncation from the five Platonic solids. To truncate literally means to cut off. Truncation thus implies the removal of some portion of a solid, actually the removal of a portion near each vertex along with the vertex itself. This can be done to the Platonic solids in such a way that the new faces are again regular polygons while the portions of the former faces that are left also form new regular polygons. For example the tetrahedron can be truncated so that the four triangles become four hexagons and the new faces are four new triangles. Five Archimedeans are thus generated. They are named simply: the truncated tetrahedron, the truncated hexahedron (cube), the truncated octahedron, the truncated dodecahedron, and the truncated icosahedron. Another subset, containing only two members, is that known as the quasi-regular polyhedra. The designation 'quasi-' implies that the solid has only two kinds of faces, each face of one kind entirely surrounded by that of the other kind. They are the cuboctahedron and the icosidodecahedron. You will find a fuller treatment of these two later on in this book (see pp. 25 and 26).

Then there are the two called the rhombicuboctahedron and the rhombicosidodecahedron. These two are sometimes named the small rhombicuboctahedron and the small rhombicosidodecahedron to distinguish them from two others called the great rhombicuboctahedron and the great rhombicosidodecahedron. If truncation is applied to the two quasi-regular solids, the cuboctahedron and the icosidodecahedron, the new faces that arise are at best rectangles and thus do not come out as regular polygons. But with some modifications these rectangles can be turned into squares. Because of this some authors refer to the great rhombicuboctahedron and the great rhombicosidodecahedron as the truncated cuboctahedron and the truncated icosidodecahedron. In this book they are named the rhombitruncated cuboctahedron and the rhombitruncated icosidodecahedron. The prefix *rhombi-* implies extra square faces (across edges) of the two quasi-regular solids. With this the designation *small* may be dropped from the names of the former two.

Finally there are two snub versions, one of the cube and one of the dodecahedron. This snub quality gives these a twisted appearance which makes each of them turn out in either of two forms—with a right- or left-handed twist. These mirror image pairs are also called enantiomorphous pairs.

If you are ambitious enough and systematic enough you can also prove to your own satisfaction that the total enumeration of thirteen is complete, that there are no more, by using the same approach here that you used for the five Platonic solids. The appropriate theorem from solid geometry that applies here states that the sum total of the face angles of any convex polyhedral angle is less than 360 degrees. After you have tried all possible combinations of regular polygons for which this theorem remains true you will come up with exactly the thirteen Archimedean solids, and two infinite families, the prism (with square side-faces) and antiprisms (with equilateral triangular side-faces).

(For further details see L. Lines, *Solid geometry*, pp. 159–67.)

The union of these two sets, the Platonic and the Archimedean solids, together with the two infinite sets of prisms and antiprisms, yield the set known as the convex uniform polyhedra. A polygon is convex if no interior angle is greater than 180 degrees. Analogously a polyhedron is convex if no dihedral angle (formed by the intersection of two faces with its vertex on or in an edge) is greater than 180 degrees. Convex is the opposite of concave, bending in on itself. A polyhedron with dimples, dents or grooves in it would be non-convex or concave. The word 'uniform' implies that all faces are regular polygons and all the vertices of the polyhedron are alike. In a uniform polyhedron the polygons around any vertex occur in the same order in every other vertex. For example in the rhombicosidodecahedron the order going around a vertex is: a triangle, a square, a pentagon, and another square. The same holds true at every vertex.

The word 'enantiomorphous' occurs frequently in the following pages. It simply means the property of being right- or left-handed, as in a pair of gloves, or in mirror image pairs. In colour arrangements, if the order around a vertex of a polyhedron is taken in clockwise rotation, the enantiomorphous arrangement will be obtained by taking the same order of colours in counter-clockwise rotation.

The following abbreviations will be used for colours: Y yellow, B blue, O orange, R red, G green, W white, T tan.

All this terminology and all these classifications will undoubtedly become clearer to you and more meaningful after you have made the models in section I, the convex uniform polyhedra.

Mathematical classification

This section may be omitted at a first reading.

A uniform polyhedron can be enclosed within a sphere, so that its axes of symmetry pass through the centre of the sphere. By central projection the edges of the polyhedron can then be made to generate a network of arcs of great circles decomposing the surface of the sphere into spherical polygons, one for each face of the polyhedron. The planes of symmetry of the solid will likewise decompose the surface of the sphere into a network of spherical triangles, four for each edge of the solid if it is a Platonic solid. These spherical triangles are called Möbius triangles, because it was Möbius (1849) who first observed this. He illustrated this fact by means of his polyhedral kaleidoscope, consisting of three mirrors forming a trihedral angle. Given certain dihedral angles between these mirrors and given an object to mark a point, the images of the object (together with the object itself) mark the vertices of the polyhedron. Another perhaps easier illustration of Möbius triangles is simply a special set of great circles inscribed with chalk on a slated globe. Some of the intersections of the great circles mark the vertices of the polyhedron. This tessellated network of spherical triangles covers the globe once. All the triangles are congruent.

In symbols one of these triangles can be described by (*pqr*) where *p*, *q*, *r* are integers and the angles of the triangle are $\frac{\pi}{p}, \frac{\pi}{q}, \frac{\pi}{r}$. Here *p*, *q*, *r* can only be 2, 3, 4, or 5. But one or more of *p*, *q*, *r* may be rational; that is, certain fractions may be used as replacements for *p*, *q*, *r*, leading to Schwarz triangles. It was Schwarz (1873) who first listed the possibilities. It has been shown that a set of Schwarz triangles covering the globe more than once but still a finite number of times is equivalent to a set of Möbius triangles. Thus Schwarz triangles may be classified as tetrahedral, octahedral or icosahedral, depending on the Möbius triangles to which they are related. (For further information see: Coxeter *et al.*,

Mathematical recreations and essays, *Regular polytopes*, and *Phil. Trans.* (1954), **246**A, 401.

These ideas are easier to visualize with the aid of models. You can make your own polyhedral kaleidoscope with three mirrors cut in the shape of circular sectors. The radius must be fairly large, twelve inches or more; the central angles of these sectors must be as follows:

for the tetrahedral kaleidoscope 54° 44′, 54° 44′, 70° 32′;

for the octahedral 35° 16′, 45°, 54° 44′;

for the icosahedral 20° 54′, 31° 43′, 37° 23′.

Interesting as it is to *play* with these mirrors, they are not always so easy to come by, nor are they completely satisfactory. So it is just as good or better to make models of the spherical triangles using the same index card or coloured tag you use in the other models. By repeating these spherical triangles a sufficient number of times you can make a model of the entire sphere as an intersecting set of great circles. In fact the colours can be worked out so that they illustrate the great circles but this calls for more work than is needed in a model all of one colour.

The tetrahedral case is the easiest to begin with. The parts are cut as shown in fig. 1. Score these parts on the radial lines, then fold them forming a model of a spherical triangle. The cementing is done using only one tab as shown. Twenty-four of these are needed and they are simply cemented to each other, flat surface to flat surface, so that the tab joint disappears between the two surfaces. You may do the work in sections. One of these sections has six spherical triangles as shown in fig. 2. The angles are $\frac{\pi}{2}, \frac{\pi}{3}, \frac{\pi}{3}$. Four of these sections complete the model.

For the octahedral case you may follow the same procedure with another set of parts shown in fig. 3. Forty-eight of these are needed, two enantiomorphous sets of twenty-four in each. You may make these any convenient size. In fact

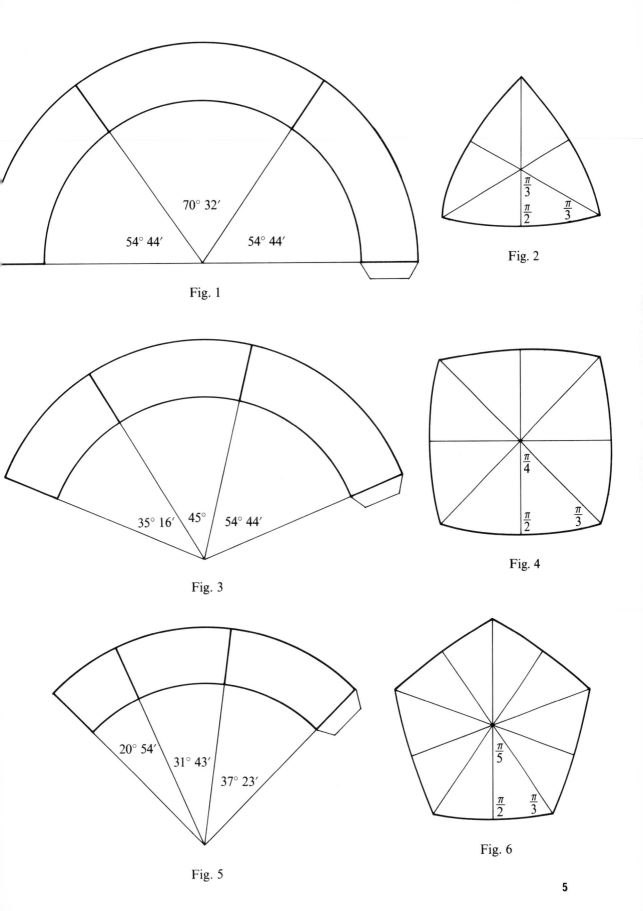

70° 32′

54° 44′ 54° 44′

Fig. 1

$\frac{\pi}{3}$

$\frac{\pi}{2}$ $\frac{\pi}{3}$

Fig. 2

35° 16′ 45° 54° 44′

Fig. 3

$\frac{\pi}{4}$

$\frac{\pi}{2}$ $\frac{\pi}{3}$

Fig. 4

20° 54′ 31° 43′ 37° 23′

Fig. 5

$\frac{\pi}{5}$

$\frac{\pi}{2}$ $\frac{\pi}{3}$

Fig. 6

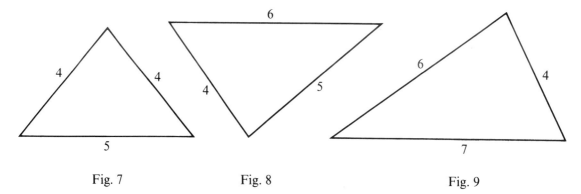

Fig. 7   Fig. 8   Fig. 9

you may make the circular band wider or narrower as you please, even leaving all the interior, which of course will bring you right to the centre of the sphere. The sections in this case begin to reveal the fact that the octahedron is the dual of the cube, since one of these sections may be the eight spherical triangles arranged as shown in fig. 4. The angles are $\frac{\pi}{2}, \frac{\pi}{3}, \frac{\pi}{4}$. Six of these sections complete the model.

The icosahedral case calls for more work because of the greater number of parts, but the procedure is still the same. It is well worth the effort it takes, because it will bring you a great deal of enlightenment. The openness of the model on the interior has great advantages. One hundred and twenty of these parts are needed, two enantiomorphous sets of sixty in each. The sections are pentagonal, ten spherical triangles to a section. The angles are $\frac{\pi}{2}, \frac{\pi}{3}, \frac{\pi}{5}$. Twelve of these sections (see fig. 6) complete the model.

There is still another way to make models illustrating Möbius triangles. It amounts to making a polyhedron whose faces are plane triangles with the same vertices as the spherical triangles. If the sides of a spherical triangle are p, q, r, namely p, q, r are the angles subtended at the centre of the sphere, the corresponding plane triangles have sides proportional to

$$\sin \tfrac{1}{2}p : \sin \tfrac{1}{2}q : \sin \tfrac{1}{2}r.$$

The three cases are shown in figs. 7, 8 and 9 and of course they call for the same number of parts respectively as the spherical triangles to which they correspond. The models may also be done by following the same sectional procedure as for the spherical cases. The numbers are approximate measures in linear units of the sides of these triangles. If you use centimetres you can get satisfactory results.

You can also get some striking colour effects by making one set of triangles all W and then the others in the usual colours. The drawings below show the respective sections and their colour tables.

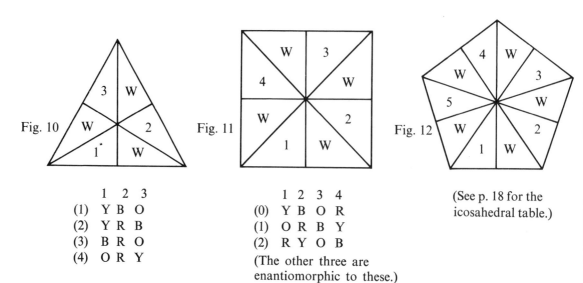

Fig. 10   Fig. 11   Fig. 12

	1	2	3
(1)	Y	B	O
(2)	Y	R	B
(3)	B	R	O
(4)	O	R	Y

	1	2	3	4
(0)	Y	B	O	R
(1)	O	R	B	Y
(2)	R	Y	O	B

(The other three are enantiomorphic to these.)

(See p. 18 for the icosahedral table.)

A. Tetrahedral

B. Octahedral

C. Icosahedral

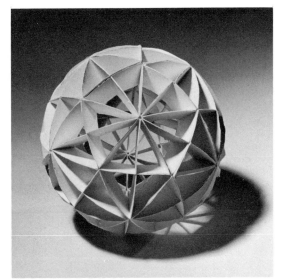

Note that the tetrahedral case has the sections numbered (1), (2), (3), (4). The tetrahedron and the truncated tetrahedron are the only uniform polyhedra that do not have their vertices distributed in diametrically opposite pairs. The (0) section in the other two cases may be taken as the north polar section. Then in the octahedral case (1) and (2) are cemented in place something like the faces of a cube. These are followed by the enantiomorphous arrangement of the same two sections, thus completing the four side-faces of the cube. The enantiomorph of (0) then completes the model. The icosahedral case is the most interesting. You may begin with the (0) section, cementing together ten triangles alternating W with one of each of the colours. Follow the icosahedral colour arrangement as shown on p. 18. Then as you complete each of the other

derived, but there is an interchange in the number of faces and vertices. Moreover the kinds of faces and vertices are such that an *n*-sided face in the original solid yields an *n*-edged vertex in the dual solid. The three polyhedrons just described thus turn out to be duals of **7**, **15** and **16**, respectively.

Once you have made these models it is a good exercise in spatial imagination to use them, especially the spherical models, to locate the faces, edges and vertices of the convex uniform polyhedra. The diagrams here may serve as a guide. The numbers designate the vertices whose images are the vertices of the polyhedron designated by the same number in the summary, p. 9. The snubs, **17** and **18**, are not indicated above. The vertices of these depend on a suitable point being chosen within the triangles. The exact

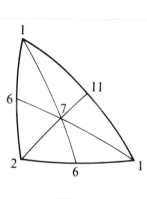

Fig. 13

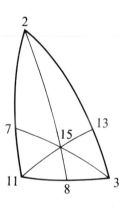

Fig. 14

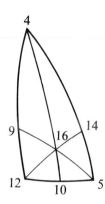

Fig. 15

sections (1), (2), (3), (4), (5), cement it to the (0) section first and then to its neighbour in dodecahedral fashion. The next set of six sections have the enantiomorphous order of colours. They are placed diametrically opposite their counterparts. You will be delighted with the pin-wheel appearance that turns up on all three of these cases. It is most pronounced in the icosahedral case. It is worth mentioning here that these three models are actually Archimedean duals. Dual solids are those which have the same number of edges as the original solids from which they are

mathematical details can be found in L. Lines, *Solid geometry*, pp. 175–84. The construction there relates the snubs to the circumscribed cube and dodecahedron, showing how to find the vertices of the snubs on the faces of the cube and dodecahedron. Then by central projection these same points can be located in the spherical triangles.

The summary that follows is an attempt to bring together the various aspects relating principally to the symbols used for each polyhedron. You need not master this material to make the

Summary of convex uniform polyhedra together with their symbols

Regular polygons:

triangle	square	pentagon
{3}	{4}	{5}
hexagon	octagon	decagon
{6}	{8}	{10}

Uniform polyhedra:

Platonic solids (regular solids)

1. tetrahedron $\quad \{3, 3\} = 3 \mid 2\ 3 = 3^3$
2. octahedron $\quad \{3, 4\} = 4 \mid 2\ 3 = 3^4$
3. hexahedron (cube) $\{4, 3\} = 3 \mid 2\ 4 = 4^3$
4. icosahedron $\quad \{3, 5\} = 5 \mid 2\ 3 = 3^5$
5. dodecahedron $\quad \{5, 3\} = 3 \mid 2\ 5 = 5^3$

Archimedean solids (semi-regular solids)

6. truncated tetrahedron $\quad t\{3, 3\} = 2\ 3 \mid 3 = 3.6^2$
7. truncated octahedron $\quad t\{3, 4\} = 2\ 4 \mid 3 = 4.6^2$
8. truncated hexahedron $\quad t\{4, 3\} = 2\ 3 \mid 4 = 3.8^2$
9. truncated icosahedron $\quad t\{3, 5\} = 2\ 5 \mid 3 = 5.6^2$
10. truncated dodecahedron $\ t\{5, 3\} = 2\ 3 \mid 5 = 3.10^2$
11. cuboctahedron (quasi-regular) $\qquad \{^3_4\} = 2 \mid 3\ 4 = (3.4)^2$
12. icosidodecahedron (quasi-regular) $\quad \{^3_5\} = 2 \mid 3\ 5 = (3.5)^2$
13. (small) rhombicuboctahedron $\qquad r\{^3_4\} = 3\ 4 \mid 2 = 3.4^3$
14. (small) rhombicosidodecahedron $\quad r\{^3_5\} = 3\ 5 \mid 2 = 3.4.5.4$
15. rhombitruncated cuboctahedron $\qquad t\{^3_4\} = 2\ 3\ 4 \mid\ = 4.6.8$
16. rhombitruncated icosidodecahedron $\ t\{^3_5\} = 2\ 3\ 5 \mid\ = 4.6.10$
17. snub cube $\qquad s\{^3_4\} = \mid 2\ 3\ 4 = 3^4, 4$
18. snub dodecahedron $\quad s(^3_5) = \mid 2\ 3\ 5 = 3^4, 5$

models in this book, but it is interesting to know that the details have been worked out. If you should ever want to undertake further investigation in this field you would have to be thoroughly acquainted with the details.

The Schläfli symbol is given first, then the symbol with dashes, "|" as used in *Uniform polyhedra*, then another symbol as used in *Mathematical models*. In the symbol $\{p, q\}$, p names the polygon that appears in the faces, q names the polygon that appears in the vertex figure. For an explanation of the dashes, see the following page. $\begin{Bmatrix} p \\ q \end{Bmatrix}$ simply names the two kinds of polygons found in the faces of the quasi-regular solids. It is an extension of the Schläfli symbol. So too are t, r, s to mean truncated, rhombic and snub respectively. Rhombic implies the existence of extra square faces. Snub implies the existence of extra triangular faces.

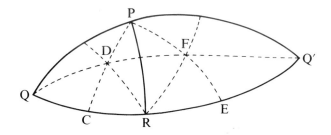

Fig. 16

The meaning of the dashes, "|", may be briefly summarized as follows: A spherical triangle PQR whose angles are $\frac{\pi}{p} \cdot \frac{\pi}{q}, \frac{\pi}{r}$ may also be named (pqr). In terms of the polyhedral kaleidoscope:

The polyhedron whose vertices are the images of P is $p|q\,r$ (or $p|rq$).

$$p|q2 = p|2q = \{q, p\}$$
$$q|p2 = q|2p = \{p, q\}$$
$$2|pq \quad = \begin{Bmatrix} p \\ q \end{Bmatrix}.$$

The polyhedron whose vertices are the images of C is $qr|p$ (or $rq|p$). C is the point of intersection of the bisector PC of the angle QPR with the opposite side QR.

$$pq|2 = r \begin{Bmatrix} p \\ q \end{Bmatrix}$$
$$2q|p = t \{p, q\}.$$

The polyhedron whose vertices are the images of D is $pqr|D$ is the incentre of the triangle PQR.

$$pqr\,| = prq| = q\,r\,p| = \text{etc.}$$
$$2pq| = \ t \ \begin{Bmatrix} p \\ q \end{Bmatrix}.$$

E and F apply only to the non-convex uniform polyhedra. They are given here to complete the summary. PE is the external bisector of the angle at P. F is the excentre.

Suppose the angle $Q'PR$ is $\frac{\pi}{p'}$; then $\frac{1}{p} + \frac{1}{p'} = 1$. If the angle PRQ is $\frac{1}{2}\pi$, then the polyhedron whose vertices are the images of E is $2q|p'$.

$$p'q|2 = r' \begin{Bmatrix} p \\ q \end{Bmatrix}$$
$$2q|p' = t'\{p, q\}.$$

The polyhedron whose vertices are the images of F is $2\,p'q|$.

$$2p'q| = \ t' \ \begin{Bmatrix} p \\ q \end{Bmatrix}.$$

The polyhedron whose vertices are the images of a suitable point within PQR is $|\,pqr$.

$$|2pq = \ s \begin{Bmatrix} p \\ q \end{Bmatrix}.$$

The symbols r' and t' stand for quasi-rhombic and quasi-truncated respectively.

The Convex Uniform Polyhedra

The Platonic and Archimedean Solids

General instructions for making models

The first thing you must do to make a model of any polyhedron is to make an accurate drawing for the required parts. For the convex polyhedra these are simply polygons of 3, 4, 5, 6, 8, and 10 sides. But you must remember that in any one polyhedron all the edges must be the same length. Hence the polygons belonging to one polyhedron must have sides of the same length. As you can see from a drawing, the decagon, for example, is very large compared to a triangle with the same edge length. You must keep this in mind when making the models and choose a suitable scale. This will be determined by how you want to use the polyhedron and where you intend to display it. In the following descriptions of the individual models a value is given for the circumradius, that is, the radius of a circumscribing sphere, in terms of a polyhedral edge length of 2 units. This will help you to determine how big the completed model will turn out to be. Doubling the radius gives you the diameter and this can be taken as an approximate value for the height of the completed model.

Once you have carefully drawn the parts, namely the required polygons, it is best to make a template. This is done by placing the drawing of the polygon over a piece of card or stiff paper. Index card stock or coloured tag is recommended. Then prick through at each vertex with a probing needle. The kind found in a biology dissecting kit serves the purpose very well. You may then draw pencil lines from hole to hole and trim the card with scissors leaving about a quarter inch border all around outside the pencil lines. This is your template.

It is now a simple and easy matter to multiply copies of the parts any number of times. This is done by placing the template on top of a number of sheets of card. It is best to staple the sheets together. Usually four, five, or six parts are your requirements at any one time and this will then be the number of sheets of card, say one of each colour needed, that you may staple together. Now again using the probing needle prick through at each vertex, using the template for this purpose as a guide. You may draw pencil lines around the edges of the template before you move it to its next position on the sheets of card. In this way you can prick as many patterns as you need or wish to do at one time. Your next step is to cut with scissors along the pencil lines left by outlining the template while the sheets of card are still held together by the staples. The sheets may have a tendency to move slightly while you are cutting, but this is not too serious because the quarter inch border all around gives you a little leeway. But once the pieces have been cut in the way just described you will get best results from here on by handling each part individually. With a sharp pointed instrument such as the point on geometry compasses, you must now score the card using a set square or ruler for a straight edge. If you plan to do a lot of model making it will pay you to take the filler from a ball-point pen and replace it with the steel point from the compasses. A mimeograph stylus also serves the purpose very well. You must draw the scoring lines to connect the needle holes. Pencil lines are not needed, since the process of scoring sufficiently outlines the shape of each part. More accurate trimming with scissors must now be done. As mentioned before you will get best results by handling each part separately. Cut directly through or into the needle holes and out again so that the quarter inch border is left as a tab. Then fold the tabs down. The scored lines make this a simple and accurate operation. These tabs will be used for cementing the parts together. Where the parts have more acute angles, you must trim the tabs again after the folding. If done before, you will find the folding more difficult. Experience will teach you how much trimming to do and how accurate it must be for best results. The rule is to leave as much as possible for the cement to hold and to remove as much as necessary so no jamming occurs at the vertices on the interior of the model.

A good household acetone cement provides

the best adhesive since it is quick drying and adheres very firmly. The procedure is to apply the cement all along one tab, then to join the tab from another part to it, to move these parts back and forth slightly to help spread the cement evenly on both parts, then to manœuvre the parts into accurate positions before the cement becomes too stiff. You will find a pair of tweezers helpful at times, especially as the work progresses and the model begins to take shape. Clamps are also helpful and even necessary on more intricate models. You can make your own clamps by taking clothes pegs of the coiled wire spring variety and turning them inside out, namely separating the parts to get the two wooden prongs reversed and then replacing the spring between them.

You will find that the method of assembly for polyhedron models suggested here will generally give you fairly rigid results, since the tabs serve as interior structural ribs along all the edges of the model. For this reason it is best to follow the general rule of leaving tabs all around on every part. It is only occasionally and in fact only rarely that you may have to depart from this procedure. This occurs only in the more complex models described later on in this book. For all the convex polyhedra it is best to leave all the tabs.

The convex uniform polyhedra are presented first. They are the easiest to make and you will find it best to begin with them. In each case the symbol designating the polyhedron is set down, followed by the kind and number of polygons comprising the facial planes of each polyhedron,

then the circumradius for the edge length of two units and finally the vertex figure. The symbol is not too important for the purposes of making the models. It belongs more to the mathematical analysis and classification of these geometrical solids (see p. 9). The number of faces and the kind of polygons appearing as faces are given by $4\{3\}+4\{6\}$; that is, the polyhedron has four triangles and four hexagons for its faces.

This information will help you to prepare the right number and the right kind of polygons in each case. In the instructions given for making individual models the word *net* is frequently used. In its context it simply means any part or parts needed in the construction of a polyhedron. Thus the word, as it is used in this book, will mean the drawing of the part or parts needed for a template. The vertex figure is also given because it is very helpful in giving you information about the order in which the polygons surround each vertex of the polyhedron. You may think of the vertex figure as the base of a pyramid, all of whose slant edges are of unit length. Or to put it another way, you may choose any vertex of a polyhedron and take note of the edges meeting at that vertex. Then the points on these edges each a unit length from the vertex of the polyhedron will be the vertices of the vertex figure. Every uniform polyhedron is characterized by its vertex figure which is a cyclic polygon (cf. Coxeter, 1954, p. 404).

For colour arrangements the map colouring principle generally gives the most striking effects; namely, polygons sharing a common side (meeting at an edge) must be of different colours.

1 The tetrahedron

The simplest of all polyhedra is the tetrahedron. It has four equilateral triangles for faces. This is the least number possible to enclose a portion of three-dimensional space. Certain properties immediately appear in it which are characteristic of the entire set of uniform polyhedra. All its faces are regular and each face shares its edges with just one other face. Also all its vertices are alike. A model of the tetrahedron can be made by using one net for the entire solid as shown. However by doing it this way, you will have all faces the same colour. So too all the convex polyhedra can be made of one net and thus of one colour. (See Cundy & Rollett, *Mathematical models*.) But if you want each face of the tetrahedron, and more generally each face of any polyhedron, to be a different colour, then you should prepare individual nets for each face that is a different polygon. For the tetrahedron all you need is one net, an equilateral triangle. Prepare four parts, each one with tabs all around as shown and each of a different colour, say Y, B, O, R. Then cement these four triangles together in the same position as shown. Now bring the remaining tabs together, cementing one pair first and letting this set firmly. Then apply cement to both remaining tabs and close the triangle down as you would close the lid on a box. The model exerts its own pressure and your fingers can do the rest along the edges until the cement is set.

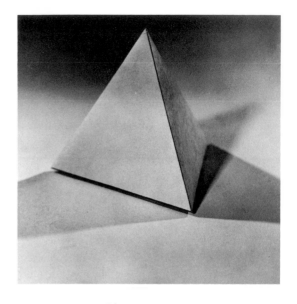

$$3\,|\,2\ 3 = \{3, 3\}$$
$$4\{3\}$$
$$\sqrt{\tfrac{3}{2}}$$

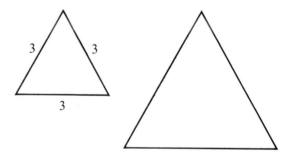

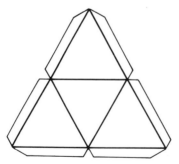

2 The octahedron

This is the polyhedron, whose faces are composed of eight equilateral triangles. Here opposite faces lie on parallel planes, so four colours serve very well. You may begin the construction of a model of this polyhedron by cementing four triangles as shown. When the remaining tabs between triangles 1 and 4 are cemented, you will have a square pyramid with triangular slant side faces but without the square base. The other tabs remain at the edges of this base. This section completes half the model.

The other half is enantiomorphous to the first half. Actually it is simpler to continue your work by cementing four triangles, one at a time, to the four tabs around the edges of the open square base. It is easy to watch the opposite faces to get the right colours. Then the tabs between adjacent triangles may be cemented and the last triangle again closed down like a lid. You may now observe that the square which showed its edges on the completion of the first section is actually only one of three such squares in the completed model. The three squares have edges on three mutually perpendicular equatorial planes. This fact is utilized in one of the non-convex uniform polyhedra **67** to be described later.

$$4\,|\,2\ 3 = \{3, 4\}$$
$$8\,\{3\}$$
$$\sqrt{2}$$

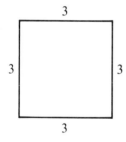

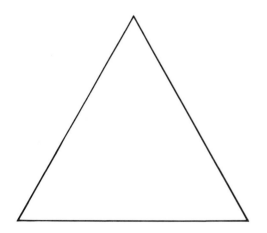

1 2 3 4
Y B O R

3 The hexahedron (cube)

The most commonly known and most widely used of all polyhedra is undoubtedly the cube, or to give it a fancier name, the hexahedron. Its six faces are all squares meeting two at each edge and three at each vertex. Since opposite faces are parallel, a simple colour arrangement is possible using three colours. You may begin the construction of a model for this polyhedron by choosing one square and then cementing four others around it as shown. Then you may cement the tabs between adjacent squares to form the four vertical edges of the cube, again with all tabs forming ribs on the interior. Finally you may add the last square, and in this case it really does answer the description which calls it the lid on a box.

$$3 \mid 2 \; 4 = \{4, 3\}$$
$$6\{4\}$$
$$\sqrt{3}$$

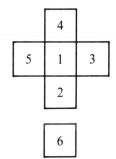

1	2	3	4	5	6
Y	B	O	B	O	Y

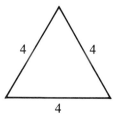

The cube may not be a very exciting poly-hedron in its own right, but it has some wonderful properties in relation to the other Platonic solids, as well as with some of the Archimedeans. A compound of five cubes can be enclosed in a dodecahedron and this makes a beautiful model. (See Cundy & Rollett, *Mathematical models*, pp. 135–6.)

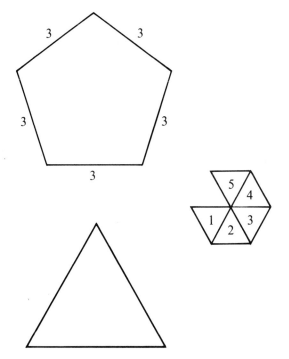

$$5 \,|\, 2 \; 3 = \{3, 5\}$$
$$20\,\{3\}$$
$$5^{\frac{1}{4}}\tau^{\frac{1}{2}}$$

4 The icosahedron

The icosahedron is one of the five Platonic solids, next in simplicity to the tetrahedron and octahedron. It shares with these the fact that all its faces are equilateral triangles. In making a model of this polyhedron there are two effective ways of arranging five colours. They can be arranged so that each of the five appears around each vertex but then opposite faces will not be of the same colour. The second arrangement has opposite faces the same but one colour repeats itself around each vertex except the two polar vertices. Both of these arrangements are very useful, because many of the uniform polyhedra to be described later in this book have icosahedral symmetry. So you will find it profitable to have two models of the icosahedron illustrating both colour arrangements for future reference. You may begin both with the same initial arrangement of five equilateral triangles as shown. These form a low pentagonal pyramid without a base. The next set of five triangles may then be cemented to the pentagonal edge as set out in the colour table. Between these the third set of five easily find their position because only one more triangle is needed to complete the ring of five at each vertex. One more set of five then completes the model.

First colour arrangement					Second colour arrangement				
1	2	3	4	5	1	2	3	4	5
Y	B	O	R	G	Y	B	O	R	G
R	G	Y	B	O	B	O	R	G	Y
	O	R	G	Y	B		G	Y	B	O	R
Y	B	O	R	G	R	G	Y	B	O

A few comments will be helpful to you in the use of the colour tables listed above. The first line may be thought of as the set of five triangles surrounding the north polar vertex of the icosahedron. The next two lines actually form an equatorial band of ten triangles, alternating with each other. The fourth line is the set of five surrounding the south polar vertex. You can also read the order of colours surrounding the other ten vertices by following the rotation of colours, starting with two adjacent colours in the first row, then proceeding into the second row and down to the third, then back to the second and finally ending in the first where you started. For example

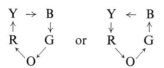

This suggests an alternative way of setting out the colour tables, a way which will be very useful for later models. This is done by listing the colours surrounding each vertex and numbering the vertices. In doing this each triangle of the icosahedron gets named three times, but the

cyclic permutation of colours is easier to follow. The alternative colour tables are:

First colour arrangement

(0) Y B O R G
(1) B Y R O G
(2) O B G R Y
(3) R O Y G B
(4) G R B Y O
(5) Y G O B R

Second colour arrangement

(0) Y B O R G
(1) Y B G O B
(2) B O Y R O
(3) O R B G R
(4) R G O Y G
(5) G Y R B Y

Only six vertices are listed, the (0) vertex being the north polar vertex, but in both cases the diametrically opposite vertices have the enantiomorphous arrangement. You can get this by reading the colour table in reverse, that is, from right to left. A little experiment with these ideas will soon make it clear to you.

$$3 \mid 2\ 5 = \{5, 3\}$$
$$12\{5\}$$
$$3\tfrac{1}{2}\tau$$

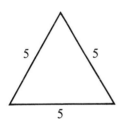

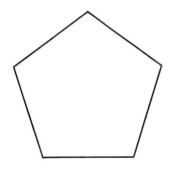

5 The dodecahedron

The dodecahedron is in some ways the most attractive of the five Platonic solids, although the icosahedron is a very close second, if second at all. The relationship of the dodecahedron to its three stellated forms, to be described later, probably gives it the advantage for first place.

A model of this polyhedron can be done in four colours in two different ways. With six colours opposite faces can be the same colour, and this arrangement carries over very well into the stellated forms mentioned above. So this arrangement is described here.

You may begin by cementing five pentagons, one of each of five colours, Y, B, O, R, G around a central pentagon which may be white, W. Once the tabs between these five coloured pentagons are cemented, half the model is completed. This half-completed dodecahedron will be mentioned again later in connection with stellated forms of the icosahedron, where it can serve as a construction cradle for other models. In such a case the tabs would be turned outward, but here you have them on the interior as usual. The other faces of the dodecahedron now easily take their positions so that opposite faces are paired according to colour.

The arrangement of four colours for the dodecahedron is shown here. It can also be done enantiomorphously. This four colour arrangement is sometimes more suitable, especially for other models having dodecahedral symmetry. Therefore it is given here for future reference.

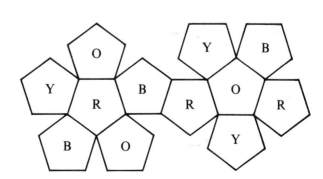

6 The truncated tetrahedron

For this polyhedron, you may obtain a very effective colour arrangement by using the same four colours for the hexagons that you used for the triangles in the tetrahedron. Then the triangles here may all be of a fifth colour. Or since these triangles are on planes opposite and parallel to the hexagon planes you may make these parallel planes the same colour. This arrangement is shown here. If you cement the parts in the order shown you will get them in their proper places. Then the remaining tabs are cemented, a pair at a time, as described before

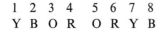

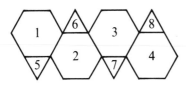

for the tetrahedron. Another way to make the model is simply to make a tetrahedral cup to begin with as shown below. This has a triangular bottom and three hexagons for sides. The tabs form ribs on the inside of the cup. Then you continue by cementing triangles and hexagons as needed. You will find it best to use a triangle as the last part and to cement one of the tabs of this triangle, first letting it set up firmly, and then closing the hole as you would close the lid on a box. This is a general procedure to follow for all models.

$$2\ 3\,|\,3 = t\{3, 3\}$$
$$4\{3\} + 4\{6\}$$
$$\sqrt{\tfrac{11}{2}}$$

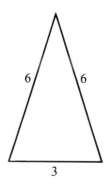

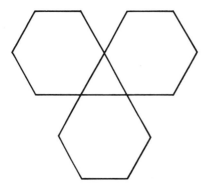

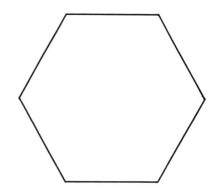

$$2\ 4\,|\,3 = t\{3, 4\}$$
$$8\{6\} + 6\{4\}$$
$$\sqrt{10}$$

7 The truncated octahedron

The way to construct a model of this polyhedron is by now becoming familiar to you, if you have done the previous ones. The colour arrangement of the hexagons follows that of the triangular faces of the octahedron, namely four pairs of opposite faces using four different colours and all the squares using a fifth colour. So you may begin here by surrounding a hexagon alternately with squares and other hexagons as shown. Then the tabs between these are cemented to form a cup, completing half the model.

Once this is done it is easy to continue cementing the other parts and watching the opposite hexagons to get the colours right. A square is added last of all. You will undoubtedly observe that complete rigidity is not achieved until the last edges have been cemented. But once this has been done, all the convex polyhedra make very rigid models.

1	2	3	4	5	6	7
Y	G	B	G	O	G	R

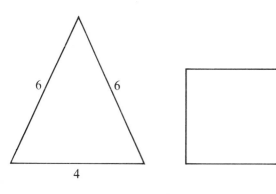

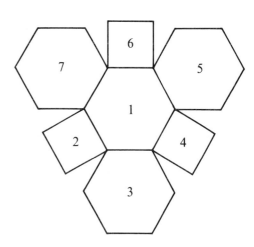

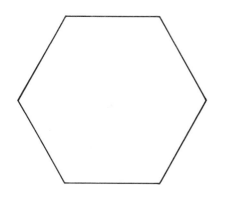

8 The truncated hexahedron (cube)

This polyhedron is a truncated cube, again not very exciting as a model, but belonging nevertheless to the set of uniform polyhedra. The colour arrangement for the octagons may follow that of the squares in a cube, leaving a fourth colour for all the triangles. You may again begin your work on this model by surrounding an octagon with triangles and other octagons as shown. The tabs on the surrounding octagons will then be cemented to each other, thus surrounding the triangles. The tabs on the triangles can be cemented as tabs on the lids of triangular holes. This is not difficult as long as you can work on the inside of the model while it is still open and incomplete. The last octagon is Y and finally four R triangles close the corners. You will see that a little more skill is being called for here to get accuracy in your work, but undoubtedly as you proceed with the work of making polyhedron models you are developing this skill.

$$2\ 3\,|\,4 = t\{4, 3\}$$
$$8\{3\} + 6\{8\}$$
$$\sqrt{(7 + 4\sqrt{2})}$$

1	2	3	4	5	6	7	8	9
Y	R	B	R	O	R	B	R	O

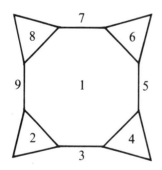

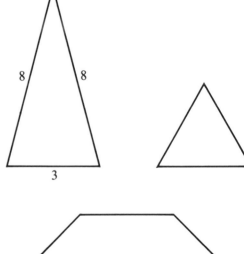

9 The truncated icosahedron

As a truncated form of the icosahedron, a model of this polyhedron may well follow the five-colour icosahedral arrangement for the hexagons and a sixth colour for all pentagons. You should have no difficulty in cementing the parts correctly if you follow the icosahedral colour table. Thus you may begin with a W pentagon and surround it with a set of five coloured hexagons Y, B, O, R, G. Then if you keep your attention on each ring of hexagons, adding the W pentagon each time at the centre of the ring, you can easily complete the next set of five rings. Each hexagon of course belongs to three rings. The completed model is very attractive with its combination of hexagon and pentagon faces.

$$2 \; 5|3 = t\{3, 5\}$$
$$20\{6\} + 12\{5\}$$
$$\sqrt{\frac{29 + 9\sqrt{5}}{2}}$$

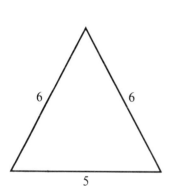

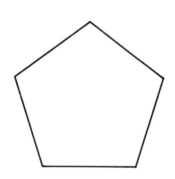

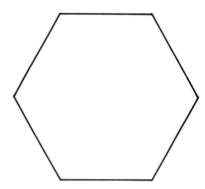

10 The truncated dodecahedron

This is the polyhedron whose facial planes are triangles and decagons. You may again use the four-colour dodecahedral arrangement for the decagons and make all the triangles a fifth colour. Around a R decagon cement in order a Y, B, G, B, G decagon alternating with O triangles. The next set of decagons is Y, R, Y, B, R; the first Y of this set adjoining the G that comes between the two B decagons in the first set. The rest of the O triangles are then cemented in place.

This polyhedron does not have a particularly pleasing shape, perhaps because the area of the decagons is very large compared with that of the triangles. For the same reason a model of this polyhedron must be made with the decagons reinforced or stiffened on the inside, say with double thickness card, otherwise these faces have a tendency to sag. On the other hand if you keep the model small, this reinforcing is not necessary.

$$2 \ 3 \,|\, 5 = t\{5, 3\}$$
$$20\{3\} + 12\{10\}$$
$$\sqrt{\frac{17 + 15\sqrt{5}}{2}}$$

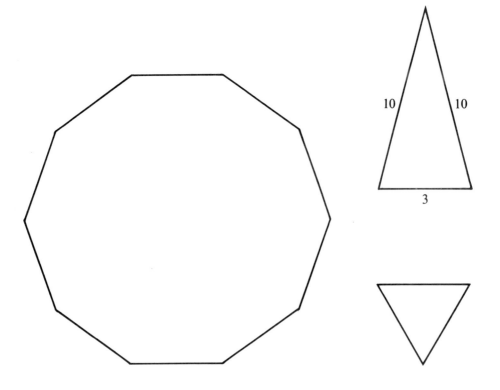

$$2|3\ 4 = \{^3_4\}$$
$$8\{3\} + 6\{4\}$$
$$2$$

11 The cuboctahedron

The name of this polyhedron suggests a close relationship to the cube and the octahedron, and indeed this is so. The six squares are on the facial planes of a cube and the eight triangles on the facial planes of an octahedron. You may later wish to make a compound of these two Platonic solids and then you will observe that the cuboctahedron is the portion of space common to the two.

To make a model of this polyhedron the three colours used for the cube may serve here for the squares and a fourth colour for all eight triangles. You may begin with a triangle and cement a square to each of its edges as shown. Then three more triangles between these squares will complete half the model, a kind of cup with a triangular bottom and squares and triangles alternating for sides. Once this section is completed you may now easily continue and get the arrangement of colours right by observing the opposite squares for the correct colour.

An important property of this polyhedron is the fact that it has two types of faces, each kind entirely surrounded by that of the other. As such it is designated quasi-regular.

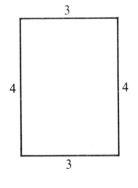

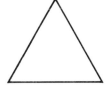

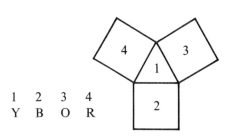

1 2 3 4
Y B O R

12 The icosidodecahedron

This polyhedron is a combinatorial solid, quasi-regular, in the same way as the cuboctahedron. It is the interior part common to the compound of an icosahedron and dodecahedron. If you limit yourself to five colours, a suitable arrangement can be worked out for a model of this polyhedron by making all the triangles Y and using the other four colours for the pentagons. This follows the four-colour arrangement for the dodecahedron.

Thus you may begin with a B pentagon and cement five Y triangles to its tabs. Next five more pentagons are cemented so each shares two of its edges with two adjacent triangles already in place around the B pentagon. The colours should be O, R, G, R, G. Another set of five triangles will then complete half the model, leaving a ring of tabs in the form of an equatorial decagon. In doing the second half of the model you may proceed by adding alternately triangles and pentagons to the equatorial edges as just described. Place an O pentagon so its vertex coincides with the vertex of the G pentagon that appears between the two R pentagons. The order, repeating the O in naming the five colours, is then: O, B, O, R, B. The last G pentagon is added as soon as some of the last five triangles are in place. The remaining triangles then complete the model. You will now notice five other sets of equatorial edges. This property is used for some of the other non-convex uniform polyhedra.

$$2 \,|\, 3 \;\; 5 = \left\{ \tfrac{3}{5} \right\}$$
$$20\{3\} + 12\{5\}$$
$$2\,\tau$$

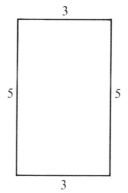

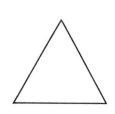

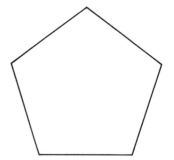

13 The rhombicuboctahedron

The name of this polyhedron again indicates its nature. The set of squares forming its faces break up into two subsets and thus the colour arrangement for these may well do the same. The triangles may then all be of another colour. To make a model of this polyhedron you may begin by making a section which forms a shallow cup having an octahedral upper edge as shown.

Next a square is cemented to each tab at the octahedral upper edge of the cup. These squares alternate in colour as set out below. In following the map colouring principle you will easily see that each R square must share an edge with a B triangle, and each Y square must share an edge with a R square. The rest of the model is then easy to do, one part at a time, and continuing with the alternating colour arrangement for the squares. This turns out to be a rather attractive model even though it is composed of only triangles and squares.

It is worth mentioning that a pseudo-rhombicuboctahedron can be formed by rotating an octagonal 'cap' of the rhombicuboctahedron through an angle of 45° relative to the rest of the solid. There then arises a solid with all vertices alike but not Archimedean. This is because the cubic and rhombic squares get mixed up.

$$3 \ 4 \mid 2 = r\{^3_4\}$$
$$8\{3\} + (12+6)\{4\}$$
$$\sqrt{(5+2\sqrt{2})}$$

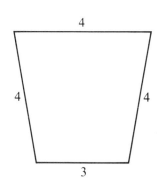

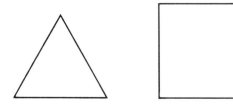

1	2	3	4	5	6	7	8	9
Y	B	R	B	R	B	R	B	R
	R	Y	R	Y	R	Y	R	Y

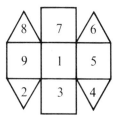

14 The rhombicosidodecahedron

This polyhedron is in some ways the most attractive of the Archimedean solids. The simplest and most suitable arrangement of colours for a model of this polyhedron is obtained by making each of the three different kinds of faces a different colour, say all triangles Y, all squares B, and all pentagons O. Then you may work around each pentagon and complete the rings with alternate triangles and squares in such a way that adjacent rings share two triangles and a square in common. You will find variations of this polyhedron turning up in the non-convex uniform polyhedra described later in this book. However, other arrangements of colour will be suggested there. They could of course also be used here, quite effectively.

$$3 \ 5 \,|\, 2 = r\{\tfrac{3}{5}\}$$
$$20\,\{3\} + 30\,\{4\} + 12\,\{5\}$$
$$\sqrt{(11 + 4\sqrt{5})}$$

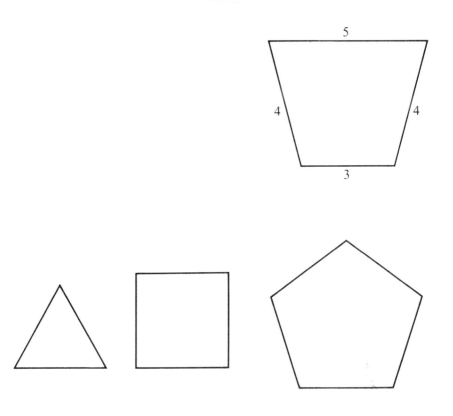

15 The rhombitruncated cuboctahedron

This polyhedron, also known as a truncated cuboctahedron, again lends itself to a simple colour arrangement. Three colours will serve in pairs for opposite octagonal faces, then all hexagons share a fourth colour and all squares a fifth colour. Thus to make a model of this polyhedron you may begin as usual with a cup shaped section as shown. Once this is completed four more octagons are cemented in place as set out in the colour table. You will then have no further difficulty in completing the model. As these models now become more intricate they also become more interesting and attractive.

$$2\ 3\ 4| = t\{^3_4\}$$
$$8\{6\} + 12\{4\} + 6\{8\}$$
$$\sqrt{(13 + 6\sqrt{2})}$$

1	2	3	4	5	6	7	8	9
Y	B	O	B	O	B	O	B	O
		R		G		R		G

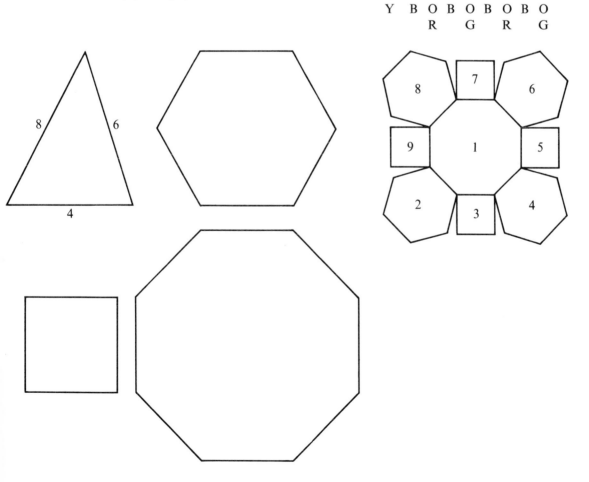

16 The rhombitruncated icosidodecahedron

This polyhedron is also called the truncated icosidodecahedron. Here too the simplest colour arrangement turns out to be the best, that is, three colours, say Y for the decagons, B for the hexagons and O for the squares. You may also use the same procedure for constructing this model as you used for the last one, working around each decagon to make a ring of alternating hexagons and squares. Again each ring shares two hexagons and a square in common with adjacent rings. This polyhedron also has its analogous forms among the non-convex uniform polyhedra to be described later. Again where the decagon plane appears in this model you must be sure that it is stiff enough so that it will not sag. So here too if you keep the model small you will automatically get a better result.

$$2\ 3\ 5\,|= t\,\{\tfrac{3}{5}\}$$
$$20\,\{6\} + 30\,\{4\} + 12\,\{10\}$$
$$\sqrt{(31 + 12\sqrt{5})}$$

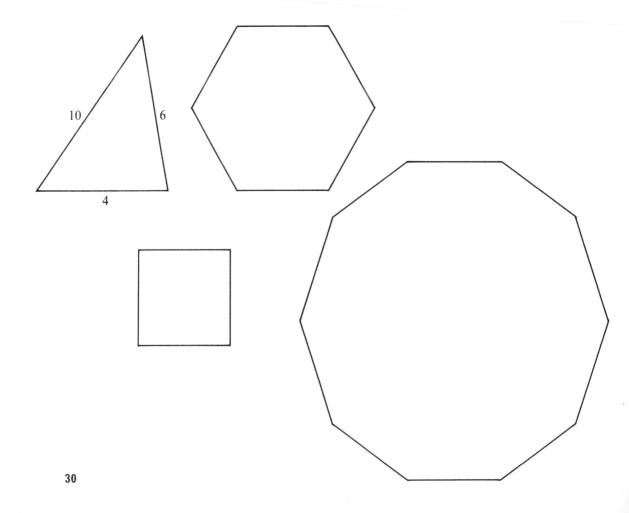

17 The snub cube

This polyhedron can be inscribed in a regular cube in such a way that its six square faces will be coplanar with those of the cube but will be in a slightly twisted position (cf. L. Lines, *Solid geometry*, p. 76). Each square is entirely surrounded by triangles accounting for twenty four of these. Then eight more triangles close the remaining spaces to complete the solid. This suggests the following colour arrangement. The squares may have three colours in opposite pairs. Each of these square faces will be entirely surrounded by triangles of the same colour. Thus the same three colours will occur in the triangles but shifted to maintain the map colouring principle. Finally the other eight triangles share a fourth colour.

To make a model of this polyhedron you may follow the colour table set out below, showing the arrangement for the first three sections. These sections must now be cemented together using R triangles as connectors between the other triangles. These three sections complete half the model. You can do the other half in the same way provided you get the squares in colour pairs opposite each other.

$$|2\ 3\ 4 = s\left\{\begin{smallmatrix}3\\4\end{smallmatrix}\right\}$$
$$(8+24)\{3\}+6\{4\}$$
$$2{\cdot}68742\ 67475$$

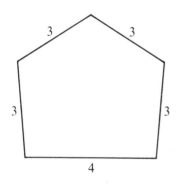

1	2	3	4	5	6
Y	B	B	B	B	R
B	O	O	O	O	R
O	Y	Y	Y	Y	R

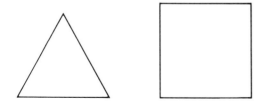

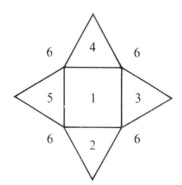

18 The snub dodecahedron

This polyhedron has the same relation to the regular dodecahedron that the snub cube has to the regular hexahedron. To get a suitable colour arrangement in a model of this polyhedron you may make all the pentagons Y. Then each of these is to be totally surrounded by triangles, each pentagon getting five triangles of the same colour. The sets of triangles however may be done in four colours. These parts are then assembled in the four-colour dodecahedral arrangement using Y triangles as connectors. The colour table shown below will help you.

This is the last of the set of convex uniform polyhedra. The non-convex uniform polyhedra are given after the set of stellations and compounds which follows.

$$| 2\ 3\ 5 = s\{^3_5\}$$
$$(20+60)\{3\}+12\{5\}$$
$$4\cdot31167\ 47491$$

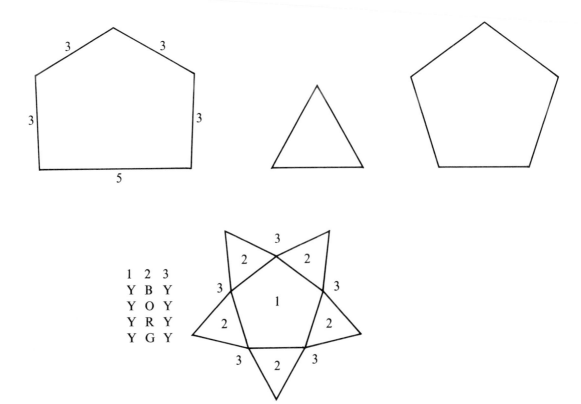

1	2	3
Y	B	Y
Y	O	Y
Y	R	Y
Y	G	Y

II Some Stellations and Compounds

Commentary on stellations and compounds of the Platonic solids

The word 'stellation' comes from the Latin word 'stella' which means 'star'. There are star polygons as well as star polyhedra. Exactly what this means is best understood by drawings and models rather than by abstract definitions. You can thus begin again with the simplest polygon, the equilateral triangle, and see what happens if you produce each line segment forming its sides. You find that no new portions of two dimensional space can be enclosed. See fig. 17. The lines will forever get farther apart. The same thing happens if you try producing the sides of a square. The lines are in parallel pairs and will never meet to enclose any portion of the plane other than the interior of the original square. See fig. 18. With the pentagon something more interesting begins to happen. The sides of the pentagon when produced will meet and enclose more space exterior to the original pentagon. It

turns out to be the well-known five-pointed star, also called the pentagram. See fig. 19. It was known to the ancients, as is evident from the fact that it was used by the Pythagorean brotherhood as a symbol of health. Similarly a hexagon leads to a six-pointed star or hexagram (really not a single polygon but a compound of two equilateral triangles). An octagon leads to an eight-pointed star or octagram; a decagon to a ten-pointed star or decagram. The pentagram, octagram, and decagram can still be considered as single polygons of five, eight, and ten sides respectively, since you can trace out their sides in a continuous movement going around the centre of the figure twice in the case of the pentagram instead of once as in the pentagon. In the figures follow the order of the numerals. In the octagram and decagram a continuous movement will take you three times around the centre. Note

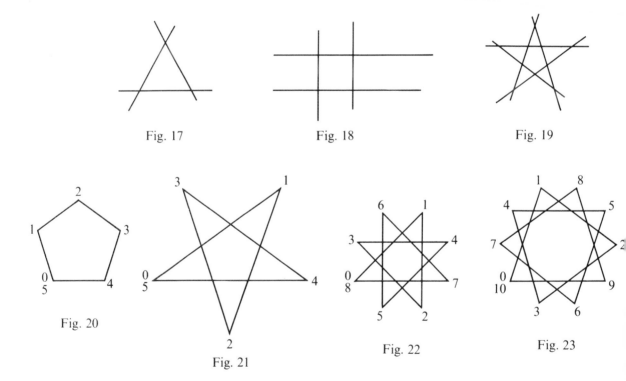

Fig. 17

Fig. 18

Fig. 19

Fig. 20

Fig. 21

Fig. 22

Fig. 23

34

that the internal points of intersection are disregarded. These facts are expressed symbolically in the fractions used for naming these star polygons: $\frac{5}{2}$, $\frac{8}{3}$, $\frac{10}{3}$. These stars can assume other shapes as well, but only these are mentioned here because only these appear again later in this book. (See Coxeter, *Introduction to Geometry*, p. 36.)

If you turn your attention now to the process applied analogously to three-dimensional space you can begin again with the simplest polyhedron, namely the tetrahedron. Instead of producing line segments you must here think of extending the facial planes indefinitely. The four planes of the tetrahedron enclose only that portion of three-dimensional space which belongs already to the original tetrahedron. The six planes of the cube come in parallel pairs, mutually perpendicular, something like the two pairs for the square, the two-dimensional analogue. But no new portions of space are enclosed. The eight facial planes of the octahedron, however, lead to something more interesting. These will enclose not only the original octahedron, but also other portions of space exterior to this octahedron. You will discover that there is actually a set of eight small tetrahedra, like cells, each sharing one of its faces with a face of the original octahedron. If you now imagine these tetrahedra added to the octahedron so that the faces they share internally melt away, leaving all the interior hollow, you have a non-convex polyhedron. But you can equally well imagine it as a set of intersecting triangular faces, larger triangles than those belonging to the faces of the small tetrahedra. These larger triangles still keep the original property of a convex polyhedron, namely each edge belongs only to two faces. The edges of course also intersect each other, but the interior points of intersection on these line segments are disregarded, just as in the case of the two-dimensional stars. Each side of the pentagram for example crosses two other sides but these points are disregarded in counting the sides. So in the stellated octahedron, you still have only eight faces, and the end points of these edges are the vertices of the polyhedron.

But now a closer examination will reveal to you that this polyhedron actually turns out to be, not one polyhedron, but a compound of two —two larger tetrahedra interpenetrating and sharing a common centre, the octahedron's centre of symmetry. Kepler discovered this solid (1619) and called it the 'stella octangula'. It also has the property that its eight vertices can be made to coincide with the eight vertices of a cube while its edges are diagonals of the square faces of the cube.

Further extension of the facial planes of the octahedron will not enclose any further space, no more cells are formed, so the stellation process terminates with only one stellated form for the octahedron.

If you turn now to the dodecahedron and produce each facial plane you will find that it leads to the formation of three distinct types of cells inside the intersecting planes. Besides the dodecahedron itself, there will be twelve pentagonal pyramids. These convert the dodecahedron into the small stellated dodecahedron. Then there will also be thirty sphenoids or wedge-shaped pieces which convert the small stellated dodecahedron into the great dodecahedron. Finally there will be twenty triangular dipyramids which convert the great dodecahedron into the great stellated dodecahedron, which more literally might be called the stellated great dodecahedron. Here the stellation process stops. Thus the dodecahedron leads to three stellated forms. Two of these were discovered by Kepler (1619), the third by Poinsot (1809).

Even more interesting now is the fact that these polyhedra are not compounds as in the case of the octahedron but distinct new polyhedra. In fact they are regular polyhedra, since in two of them the faces are in each case a set of twelve intersecting pentagrams, in the third a set of twelve intersecting pentagons. It was the mathematician Cauchy (1811) who pointed out that these are in fact stellations of the dodecahedron and that these three together with the great icosahedron which is a stellation of the regular icosahedron are the only regular stellated forms possible. So to the five regular solids of the ancient world modern mathematics has added the four regular star polyhedra, whose facial planes are regular polygons or star polygons. These faces still meet at the edges by twos but they intersect each other before they do so. Note that internal lines of intersection are disregarded. The models will clearly demonstrate these facts.

For the purposes of constructing these models it is best to become acquainted with the stellation pattern as it is found in one of the facial planes— any one, because it is the same in all. For the octahedron this is a triangle within a triangle, the inner one with its vertices at the midpoints of the sides of the outer one. See fig. 24. The inner triangle is a face of the original octahedron, the outer one, of the stella octangula. For the dodecahedron a star polygon within a star polygon will give the pattern. See fig. 25. The numbering reveals which parts form the exterior portions of the facial planes. From these you can derive the nets required for the construction of the models.

In the following pages the lightly shaded portion of each facial plane indicates the portion which is visible on the top side of the polyhedron, and the darker shading, the portion visible on the underside of the facial plane. It is from the shaded portions that the nets are derived for constructing the models.

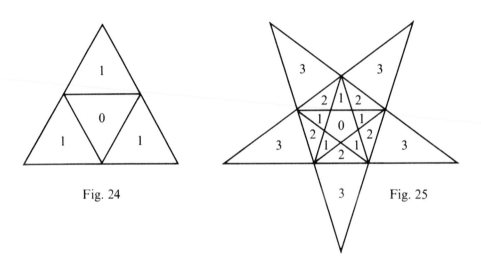

Fig. 24   Fig. 25

19 The stellated octahedron (Kepler's stella octangula)

The octahedron has only one stellated form. It turns out to be a compound of two tetrahedra. To make a model of this polyhedron all you need for a net is an equilateral triangle. The colour arrangement for the first four trihedral pyramids is given below. These parts each have triangular edges and they are cemented to each other in the same way as the octahedron itself is assembled. The pyramids here may be handled as if they were faces. You must see to it that each of the facial planes keep their own colour. You will also find parallel planes having the same colour.

The other four pyramids are the enantiomorphs. You may get these by interchanging columns 1 and 3.

Simple as this polyhedron is, it is yet very attractive.

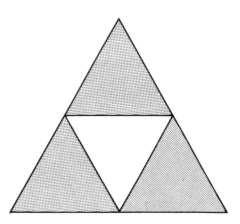

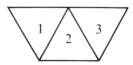

	1	2	3
(1)	O	B	Y
(2)	Y	O	R
(3)	R	B	Y
(4)	B	R	O

20 The small stellated dodecahedron

This polyhedron is one of the four Kepler–Poinsot solids. All you need for a net is one isosceles triangle with angles of 72, 36 and 36 degrees. This is the triangle found at one star arm of the pentagram, the five-pointed star. Five of these triangles are cemented together as shown below for each vertex part and the colour arrangement is given in the colour table.

The five W triangles must each be cemented to the (0) vertex part. If you remember that this colour arrangement gives you each star plane of the same colour, you will see the W star completed and two arms of each of the other star planes taking shape. You will find it more interesting to cement the parts as you complete them. The next six vertex parts are enantiomorphs and again it is best to place each one in position as you complete it. Each of these is placed diametrically opposite its counterpart.

The method of construction described here is one which gives you a completely hollow model. This may cause it not to be completely rigid. Each vertex part is slightly deformable because it is in the shape of a pentagonal pyramid without the pentagonal base. You could therefore cement the pyramids to the faces of a dodecahedron, but you will find that this procedure does not give you a neat and satisfactory result. You can get a good result with a small hollow model. Also if you apply the cement carefully along the full length of the tabs, then add another drop of cement at the concave (false) vertices around the base of the vertex parts, you will find the result to be satisfactory. On the other hand your own ingenuity may also suggest other suitable ways of obtaining the desired rigidity.

$$5\,|\,2\,\tfrac{5}{2} = \{\tfrac{5}{2}, 5\}$$
$$12\{\tfrac{5}{2}\}$$
$$5\tfrac{1}{4}\,\tau - \tfrac{1}{2}$$

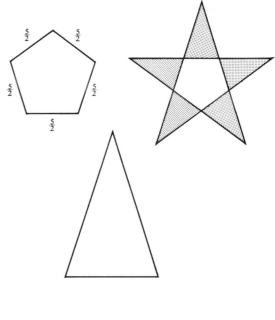

	1	2	3	4	5
(0)	Y	B	O	R	G
(1)	W	G	O	R	B
(2)	W	Y	R	G	O
(3)	W	B	G	Y	R
(4)	W	O	Y	B	G
(5)	W	R	B	O	Y

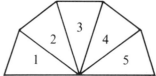

21 The great dodecahedron

This polyhedron is composed of twelve intersecting pentagon planes. When the model is made in six colours, it readily gives the appearance of a solid star embossed on a pentagon plane, but each star shares each of its arms with an adjacent star. The net for making a model is simply one isosceles triangle with angles of 36, 36, and 108 degrees. The simplest procedure for assembly is to make twenty trihedral dimples and to cement them together, very similar to the way the twenty triangles of an icosahedron are joined. The arrangement and colour table for the parts are set out below.

Triangle 5 is cemented to triangle 2. This completes half the model. The rest of the parts are the enantiomorphs, and they are placed diametrically opposite their counterparts.

$$\tfrac{5}{2}|2\ 5 = \{5, \tfrac{5}{2}\}$$
$$12\{5\}$$
$$5^{\frac{1}{4}}\,\tau^{\frac{1}{2}}$$

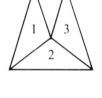

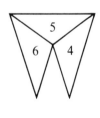

	1	2	3		4	5	6
(1)	Y	W	G	(6)	G	O	Y
(2)	B	W	Y	(7)	Y	R	B
(3)	O	W	B	(8)	B	G	O
(4)	R	W	O	(9)	O	Y	R
(5)	G	W	R	(10)	R	B	G

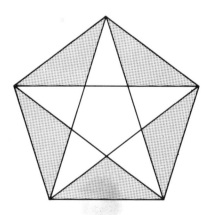

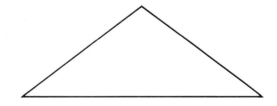

22 The great stellated dodecahedron

This is the final stellation of the regular dodecahedron. A model of this polyhedron can be made by cementing triangular pyramids to the faces of an icosahedron, but this is not recommended since it does not give a neat and satisfactory result. It is not too difficult to make the model completely hollow inside and still rigid, since the triangular pyramids are not easily deformable even with their bases missing. The net here is simply the 36, 72 isosceles triangle, one of the star arms. Cementing is done as shown below together with the colour table.

The first five parts, 1 2 3, are joined in a ring with 1 on the outer edge, which becomes pentagonal. Then the W of the other parts, 4 5 6, is cemented to 1. You will notice in doing this that the star arms take their positions so that facial planes are the same colour. The remaining parts are enantiomorphous and are placed diametrically opposite their counterparts. This model has great decorative possibilities.

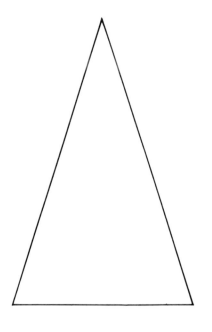

$$3\,|\,2\,\tfrac{5}{2} = \{\tfrac{5}{2}, 3\}$$
$$12\{\tfrac{5}{2}\}$$
$$3\tfrac{1}{2}\,\tau^{-1}$$

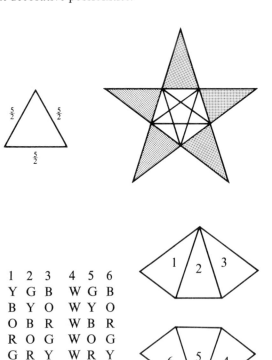

1	2	3	4	5	6
Y	G	B	W	G	B
B	Y	O	W	Y	O
O	B	R	W	B	R
R	O	G	W	O	G
G	R	Y	W	R	Y

40

Commentary on the stellated icosahedron

The meaning of stellation should now be making itself apparent to you. Some stellations are compounds. So far you have seen only one case of this, the stellated octahedron. But more are now found in the case of the icosahedron. In the case of the dodecahedron all three of its stellations turn out to be genuinely new polyhedra, in fact they all classify as regular.

The icosahedron has twenty faces and if all twenty facial planes are extended indefinitely you may well imagine, or more likely you may well fail to imagine, the multitude of cells enclosed within these intersecting planes. It is a fact that stellations of the icosahedron may all be derived from the cells enclosed within these planes. Besides the icosahedron itself you will find

$$20 + 30 + 60 + 20 + 60 + 120 + 12 + 30 + 60 + 60$$

cells of ten different shapes and sizes. The great icosahedron is composed of all but the last sixty pieces. In making models of these stellated forms, and this holds for the octahedral and dodecahedral as well, you can make these cells first, once you have worked out their nets, and then you can cement the cells to a polyhedron base or to each other. But in practice this does not give a satisfactory result and would be extremely tedious. However, acquaintance with the cell forms is very helpful in working out nets that are practical. In the following descriptions you will find these nets given for you. Once you have done some of these, or certainly by the time you have done all of them, you will be able to find many more on your own. The nets given here are not necessarily the only possible ones or the best ones. They are merely those actually used in the construction of the models pictured in the photographs.

Several compounds occur among the stellations of the icosahedron. There is a compound of five octahedra, a compound of five tetrahedra in two forms, enantiomorphs, and a compound of ten tetrahedra. Surely this would have greatly delighted the mind of Plato had he known of it.

After these and some others were discovered the question naturally presented itself: how many stellated forms are possible? In 1900 M. Brückner published a classic work on polyhedra entitled: *Vielecke und Vielflache*, in which he presented a number of new stellations of the icosahedron. Several more are due to A. H. Wheeler (1924). In 1938 H. S. M. Coxeter in conjunction with P. Du Val, H. T. Flather and J. F. Petrie gave the question a systematic investigation. By applying a few restrictive rules suggested by J. C. P. Miller to determine what forms shall be considered properly significant and distinctive, Coxeter arrived at a total enumeration of fifty-nine; thirty-two with full icosahedral symmetry and twenty-seven enantiomorphous forms with an attractively twisted appearance. Coxeter's work, *The fifty-nine icosahedra* is available from the University of Toronto Press.

The stellation pattern for the icosahedron is very interesting. It is most easily obtained by drawing one large equilateral triangle, one of the faces of the great icosahedron. On each side of this triangle you may locate two points dividing the sides of the triangle in the golden ratio. The symbol sometimes used is τ. $\tau = 1 \cdot 618$ approximately, or $\tau = \frac{1}{2}(\sqrt{5} + 1)$. The Fibonacci series is very useful here:

$$0 \quad 1 \quad 1 \quad 2 \quad 3 \quad 5 \quad 8 \quad 13 \quad 21 \quad 34 \ldots$$

The ratio of two consecutive members of this series approaches the golden ratio as a limit. With a ruler marked in sixteenths of an inch the measures $\frac{3 \cdot 4}{4}$, $\frac{2 \cdot 1}{4}$, $\frac{1 \cdot 3}{4}$, $\frac{8}{4}$ are very useful. Lines radiating from these points will give you the pattern. See fig. 26.

The colour arrangement shown in fig. 27 can be used to great advantage in every one of the stellations of the icosahedron. Fundamentally it is the icosahedral arrangement which uses five colours in such a way that each of the five is found at each vertex but in a different order from one vertex to the next. Six vertices are laid out and numbered. The other six have the enantiomorphous arrangement. This figure amounts to a colour table and for that reason it can be used again and again for all polyhedra having the icosahedral arrangement and symmetry. It is the same as the first arrangement, given on p. 18.

Many of the icosahedral stellations have a very marked dodecahedral symmetry, a fact which you may find surprising. The explanation lies in the principle of duality. The icosahedron and the dodecahedron form a dual pair; so too, the octahedron and the cube. The tetrahedron is self reciprocal, namely its dual is another tetrahedron. (See Cundy & Rollett, *Mathematical models*, p. 116.)

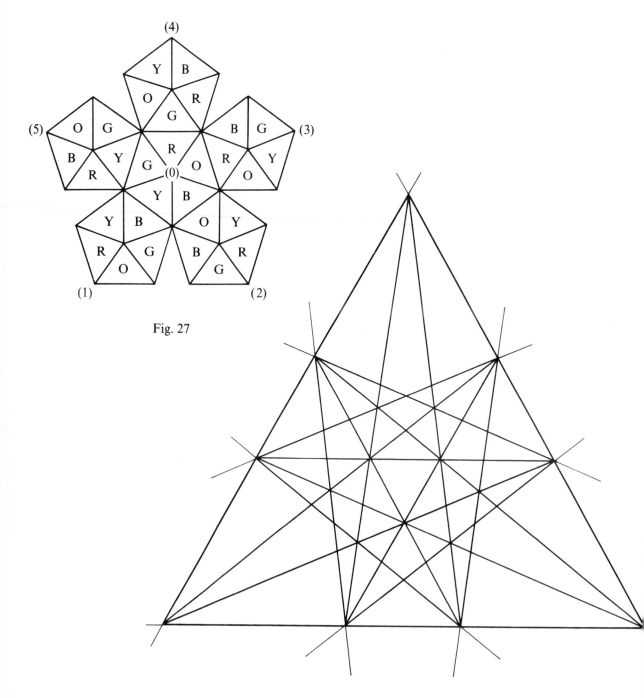

Fig. 27

Fig. 26

23 Compound of five octahedra

This polyhedron has two equilateral triangles on one facial plane, as shown in the first figure. To construct the model you may first make thirty copies of the net shown, six of each of the five colours. First assemble each of these in the form of a pyramid but without the rhombic base. Each of these will be a vertex of some octahedron of the compound. Then take a set of five, one of each colour, and cement them in the form of a ring, following the (0) arrangement of colours. Between the extending arms of this ring a second set of five vertices is cemented, but their orientation is such that the short slant edges of each pyramid continue on a line with the grooved edge between vertices of the central ring. This means that the grooved edge and the short slant edge form a straight line, part of the edge of an octahedron of the compound. If you remember to keep the basic octahedral shapes in mind you will see them begin to develop, and the colour will then help you to proceed correctly. You can in fact find the rest of the icosahedral rings beginning to appear, so this may also help you. Once you have done this much the rest is not hard to follow. This hollow model is not completely rigid, but if built on the scale indicated by the net, it will prove satisfactory. Certainly it is better than trying to add parts to a basic octahedron.

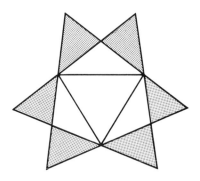

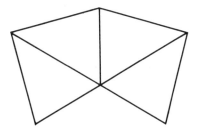

24 Compound of five tetrahedra

This polyhedron is unusually attractive because of its twisted appearance. To make a model of this compound all you need is twenty copies of the net shown below, four of each of the five colours. First make trihedral vertices with the bottom edges looking rather jagged. If you begin by making a ring of five vertices cemented together with the edges that are marked 'A' on one coinciding with the same edges of another you will find the points at the jagged end forming a dimple in the centre of the ring. Once you have built this much of the model you will easily be able to find the right positions for the other vertex parts. The colour arrangement here makes each tetrahedron entirely of one colour. The centre points of each dimple are actually the vertex points of the interior icosahedron, which of course is not being constructed. But the arrangement of colours in each dimple is the icosahedral arrangement. The method of assembly suggested here is perhaps a bit difficult to execute, because all the jagged edges and points fit into three different and adjacent dimples. The secret is to give your attention to cementing the tabs at one edge at a time, always beginning with the long edge and then working out into the dimples. The last vertex part will call for considerable skill and patience. You may find it better not to cement this part as a pyramid, but to leave one section open till last. This makes a very rigid model. It is well worth the trouble it takes to make it as suggested here. The photograph proves it can be done.

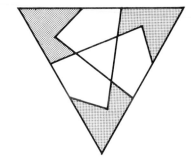

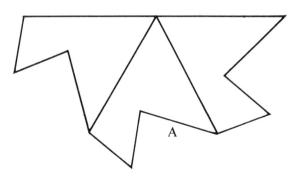

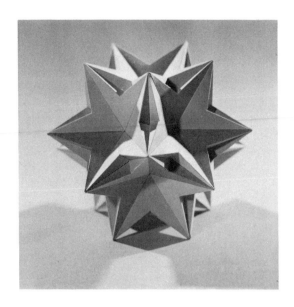

25 Compound of ten tetrahedra

This polyhedron is a combination of the two enantiomorphous forms of the compound of five tetrahedra. A model may be made by using the nets shown. If you cut the leftside arm to the centre, the triangular wing can be turned down. Then another wing as shown, but of another colour, can be cemented in place, enantiomorphous to the one you folded down. The one tab on this part which is left at its shortest side can be folded up and cemented to the under surface of the other wing. These two parts form half the grooved or cupped portion found between the sections: one of which is shown below. Their remaining tabs are used to cement these sections together. Colours for the (0) section are shown. Twelve sections are needed for the complete model. This makes a very rigid and very attractive model.

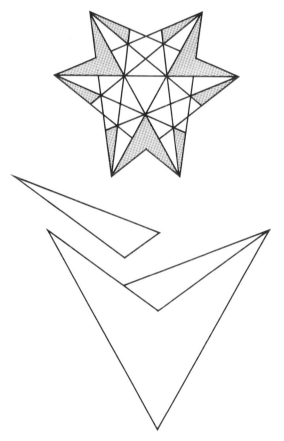

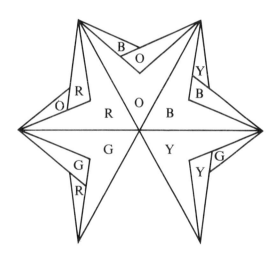

26 Triakis icosahedron

This model is the first stellated form of the icosahedron. It can be assembled from twenty parts like the net shown. These are low pyramids without their triangular base. You can get a suitable arrangement of colours using the icosahedral colour table (fig. 27, p. 42), but if you make each part as one colour you will not have the facial planes the same colour. To get this each triangle of the figure would have to be done as a separate part. You may wish to work this out for yourself.

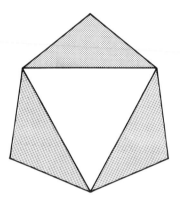

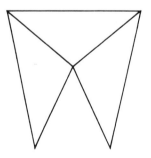

27 Second stellation of the icosahedron

This is a very beautiful model with twelve long spikes of pentahedral symmetry radiating from the dimples of the compound of ten tetrahedra. In the nets shown here you can again work out a symmetrical colour pattern that is very suitable without having facial planes the same colour. The figure shows one face of a pentahedral spike, the two smaller triangles attached to it being folded up and cemented to form a groove. Five of these make one spike with the grooves radiating away from it at its base. The other net is like the part used in the compound of ten tetrahedra except that here it is all of one colour. It serves as a connector for the spikes and is cemented to the grooves, namely between them. Again you may wish to see what colour arrangements you can discover for yourself from the icosahedral colour table.

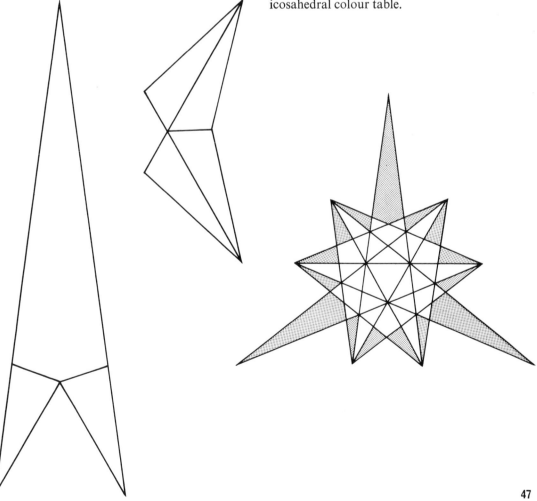

28 Third stellation of the icosahedron

This very simple polyhedron is a deltahedron. There is a whole family of deltahedra, whose common feature is that they all have faces which are equilateral triangles. (See Cundy and Rollett, *Mathematical models*, pp. 142–4.) Here three equilateral triangles are to be found in each facial plane. It has the edges and even the appearance of a dodecahedron, but it is actually one of the stellations of the icosahedron. It may be imagined as a dodecahedron with the pentagons removed, and then pentahedral dimples whose faces are equilateral triangles replacing the pentagons. This suggests a simple method for making a model of this polyhedron. You may follow the icosahedral colour arrangement exactly as shown on p. 42. The pentahedral dimples each form a section, and these sections are then cemented together in dodecahedral fashion. The colour arrangement suggested here will give you the three equilateral triangles on each facial plane the same colour. You may therefore consider this polyhedron as a three-dimensional analogue of the colour table itself.

Many of the other stellated icosahedra exhibit portions of these equilateral triangles in their facial planes. This fact makes this model particularly useful for its relationship to the other forms. In some of the models that now follow, you will see how a section of this one can serve as a structural mould or cradle. Such a section is shown here. The centre can be depressed or elevated before cementing, depending on how it is to be used.

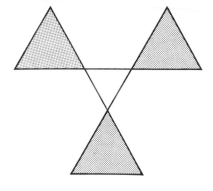

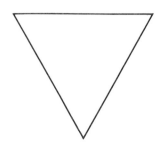

29 Fourth stellation of the icosahedron

The stellation process applied to the icosahedron leads to ten different types of basic cells, as mentioned above. One of these, whose net is shown here, can be built into a polyhedron which has these cells only vertex connected. It is equivalent to a stick model of the regular dodecahedron, but the cells here are used in place of the sticks. The method of assembly is as follows. First make a cradle or mould to hold the cells. This is half the deltahedron **28**, the interior serving as the mould. You will find it best to cut holes into the vertices of the cradle so the cells will not adhere to it when the cementing is done. Once you have this cradle ready, make five cells, one of each colour, and place them in a ring in the bottom of the cradle. The cells have only their vertices touching each other. A drop of cement on these vertices will do the trick. Let this ring set for a time. Meanwhile prepare more cells. When these are ready they can be placed with one vertex of each resting on the two vertices already cemented, while the long edges follow the slant edges of the cradle. Again apply a drop of cement. But now this must be allowed to set up firmly. After about two hours it can be moved and turned in the cradle so that other cells can be placed in the bottom to be cemented. In this way the rings of cells are completed one after the other. A little patience and a steady hand should give you a very attractive model. The colours can be worked out so that diametrically opposite cells are the same colour. You can then observe that six cells of one colour have the long edges on the faces of a cube, if you imagine a cube enclosing the dodecahedral form.

This is the first example showing dark and light shading to illustrate the top and the underside of the same facial plane.

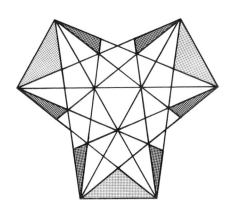

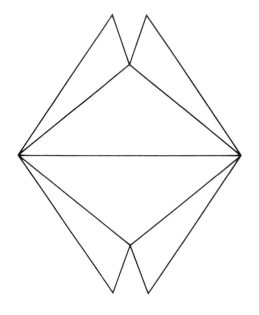

30 Fifth stellation of the icosahedron

Many of the stellations of the icosahedron appear outwardly very similar to the great icosahedron to be presented later. One of these whose nets are shown here makes a particularly pleasing model, because it is again an example of a polyhedron whose parts or cells are only vertex connected. The triangles that appear in the nets are the same as those of the great icosahedron. For that reason it may be better for you to turn to the great icosahedron first and to make that model before attempting this one. The same technique of cementing the vertex parts holds here and the same colour table is used here. In both cases it will give you a model with each facial plane the same colour. This polyhedron differs from the great icosahedron, however, in that a vertex part turns out here to be a complete polyhedron, actually an enclosed set of cells, in the form of an intricate star pyramid with convex and concave side faces and a pentagram shaped base coming to a blunt point at its centre. You should easily recognize this base as equivalent to the dimples in the compound of ten tetrahedra. Twelve of these star pyramids make the complete model. They are vertex connected at the five vertices surrounding the base in such a way that small chinks appear through which the interior can be seen. The bases of the star pyramids form the interior. These star pyramids are easy to assemble. First you must make the five that correspond to the numbered vertex parts (1), (2), (3), (4), (5). The (0) part is done last in this case. When these five pyramids are made they are laid out in a ring on a pallet shown on p. 51. This pallet may be made out of a heavy piece of cardboard. It is the regular pentagram but it is best to cut holes in the card at the places indicated by the small circles. It is at these points that the cementing of the sections is to be done. They are laid down so two edges coincide with the two sides of a star arm. When this is done you will find two vertices coming in contact between each

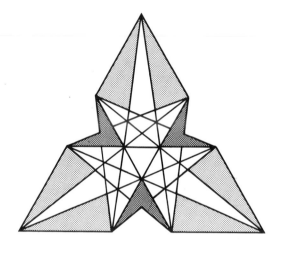

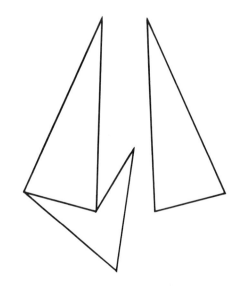

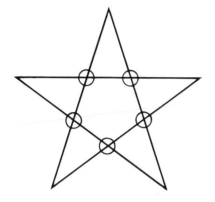

section and adjacent sections, one on the plane of the pallet and the other directly above it. A drop of cement at these vertex contact points will do the trick. After about 15 minutes another drop of cement will help to give added strength. After an hour or two you can carefully move the whole ring off the pallet. The surprise is that it holds together so that you can now turn it over. Then if you have done a careful job the (0) section will fit nicely on top with the five corners at its base making good contact with the double vertices already cemented. So on this (0) section you can apply the cement to all five corners at once. This completes half the model. The second half is enantiomorphous. Getting the sections properly oriented is a bit tricky. You may have to puzzle over this for a while, but if you remember that the arrangement is such that each facial plane is of the same colour, passing from one section to another, you should get it after a little trial and error. Once the second ring is done the first half can immediately be cemented to it before turning the second ring over. Then one more section completes the model.

31 Sixth stellation of the icosahedron

The parts shown here are the nets for another stellated form of the icosahedron. These are easy to recognize as the twelve long spikes, radiating this time from the dimples of the deltahedron, **28**. You may begin by assembling a ring of the lower parts in the usual icosahedral arrangement of colours. Then assemble a spike in the same arrangement. It is then a simple matter to insert the base of the spike into the hole at the middle of the ring and to cement the tabs there one at a time with the aid of clamps to hold them till the cement is set. The twelve sections are then assembled in the usual dodecahedral manner. A little experimentation will soon show you the correct orientation of parts so as to get the facial planes each in its own colour. This makes a simple and sturdy model which at the same time is also very attractive.

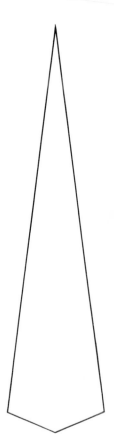

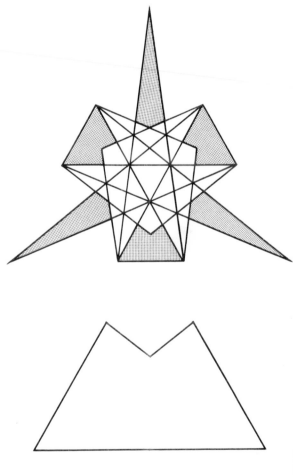

32 Seventh stellation of the icosahedron

The net shown below may be used to make twenty parts, each a hexahedral short spike. These parts can then be cemented together to form another stellated icosahedron. Four of each of the five colours may be used to make a suitable model, although in this way it will not have facial planes of one colour. You may begin with a ring of five parts, cemented so as to have the acute angles at the bottom of the parts all pointing inward to the centre of the ring. Once this is done the rest is not hard to complete by doing each ring in the icosahedral arrangement of colours.

This model is even more attractive when it is made showing each facial plane in its own colour. See if you can work out this colour arrangement for yourself.

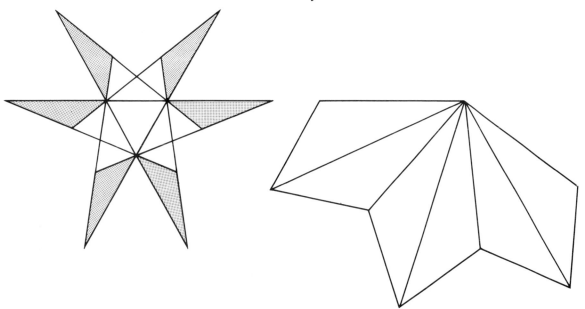

33 Eighth stellation of the icosahedron

The net shown here can be used to make a model of a stellated icosahedron that is very similar to the great icosahedron. In fact you may imagine this polyhedron as being formed by the removal of the wedge-shaped cells found around the base of the vertex parts of the great icosahedron. That is why the net shown here is slightly different at the lower end of the triangle from the triangle used for the great icosahedron. Also only one such triangle is shown here because the one net can serve for all the parts, sixty as shown and another sixty enantiomorphous to these. You may then follow the same paired arrangement of parts as you find given in the colour table for the great icosahedron. The sections are a bit more difficult to cement to one another because of the deeper depressions at the bases of these parts. But as you near the end of your work you can also change your method of assembly, doing these concave edged first and then cementing the convex parts where you can more easily pinch the parts together from the outside.

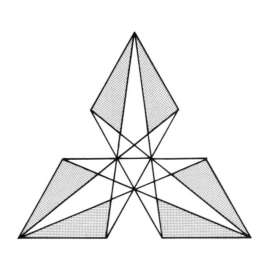

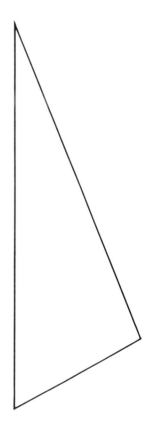

34 Ninth stellation of the icosahedron

The net shown here is again easy to recognize as part of the long spikes, only here it is slightly longer because the model is made up of twelve such spikes and nothing else. So the spikes are to be assembled in the usual pentahedral form following the usual icosahedral arrangement of the colours. Then these parts are cemented to one another along the tabs at the base. This model can be made so each facial plane has its own colour but you may not find this arrangement immediately. A little searching should reveal it to you. Here also the last spike is a bit difficult to cement in place since you cannot reach the last tabs from the outside. But patience will give you a lovely model.

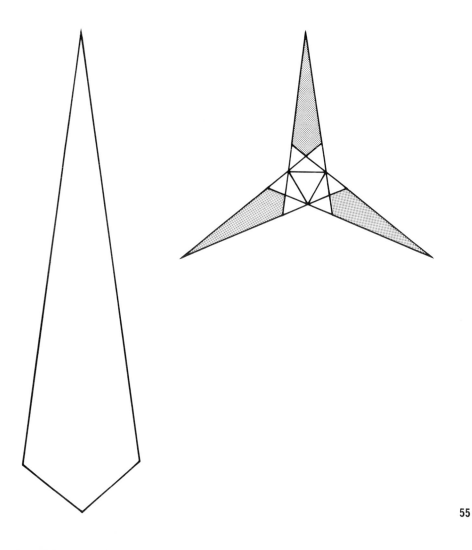

35 Tenth stellation of the icosahedron

The spike shown opposite is the net for one of the basic cells arising from the stellation process applied to the icosahedron. It is the only cell that comes in two enantiomorphous forms, sixty of one form and sixty of the other, and it is simply a short tetrahedral spike. But the interesting thing about it is that a set of sixty can actually be built into a fantastically delicate yet surprisingly stable model. Basically these spikes form all the edges of the compound of five tetrahedra, meeting by three's at each sharp pointed vertex and by two's at one of the base vertices. At first sight it looks quite impossible to make a model of this polyhedron which shows mostly empty space. But with other parts, which are later discarded, serving as structural supports, it becomes possible to make a model. These structural parts are shown opposite. Actually each of these nets can be used to construct two other stellated icosahedra in their own right. These will be described later. But the following description will help you assemble what may well be called the most unbelievable of all models. You may begin by making sixty spikes, twelve of each of the five colours, using the spike as a net. The colour arrangement in the completed model will simply be that of the compound of five tetrahedra. You will then need only five casings and five cores; their colour is not important since they are only structural and their names here indicate the purpose they serve. Turn all the tabs out on the casing, but cement only the two longer ones to form the trihedral angle, congruent to a vertex angle of the regular tetrahedron. The three quadrilaterals at the bottom should also have tabs but no cementing is done; these quadrilaterals serve as little trap doors to be opened and closed when inserting and removing three spikes for cementing and a core to keep the spikes in position. You will notice that the figures for the casing and the core have an arc mark at the vertices. This means that you must cut these parts away so as not to interfere with

the cementing. The core is made with all tabs inside as usual for any convex polyhedron. You will no doubt recognize this part as the spike found in other stellations of the icosahedron. Once all these parts are ready you may proceed as follows. Take three spikes of the same colour and insert them into the casing so that they fit snugly into the three corners. If you now insert the core so that it is centrally located inside the casing and then adjust the spikes so that their blunt ends are in the same facial planes as the facial planes on the blunt end of the casing and core, you will be able to close the little trap doors of the casing so that they fit perfectly. If the tabs on the trap doors have been turned outward then three clamps will hold everything securely while you apply a drop of cement at the point where the vertices of the spikes coincide. Repeat this with the next set of three spikes of a second colour in another casing and with another core, and so on until you have five sets of three. After about an hour you will find that the spikes may be removed from the casing and the one drop of cement will prove to be strong enough to hold the spikes to the form given to them by the casing and the core. Now you must place these five sets of triplet-spiked parts on a special platform or cradle. This is made of one section of the deltahedron **28** turned over so the dimple becomes a low pentagonal pyramid to which at the lower portions of its slant sides five of the **29** parts are cemented. The five parts may be cemented because they need never be removed, but it is advisable to cut away the obtuse vertices of **29** where cementing of cells is to be done. This is your platform or cradle. You will now find that the five sets of triplets may be placed on this platform so that the vertices point down and coincide with the vertices of the pentagonal base of the platform while two other vertices at the blunt ends of two adjoining parts will just touch. A little experimenting with the parts will make this more clear than words can do. Once the parts are so placed you can again add a drop of cement to the points of contact. After two hours or so this whole section or ring will be sufficiently rigid so that it may be moved and turned. As you complete other triplets they are placed on the platform and each ring is thus completed while the model is turned each time to accommodate the new parts. Admittedly you will have to exercise a lot of patience and a very steady hand. Also some side supports are helpful but not absolutely necessary—the dodecahedron shell is adequate and you may be able to devise other ingenious ways to get pressure in one direction or another, to achieve contact at the proper points. The photograph shows that it can be done, and it has been done in the manner described above.

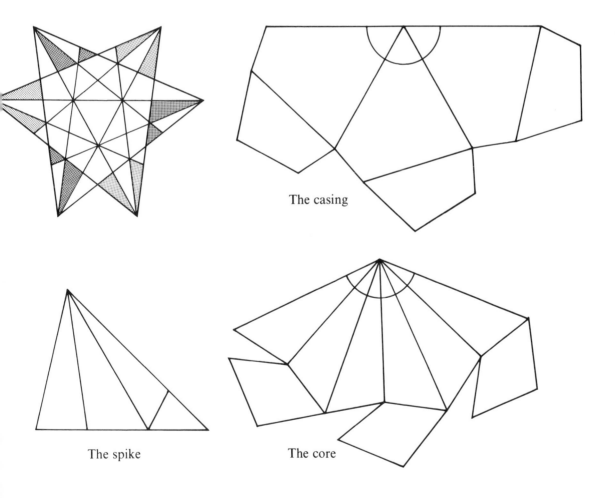

The casing

The spike   The core

36 Eleventh stellation of the icosahedron

A model of **36** can be made using the same platform and technique as for **35**. The net to use is the figure called 'the casing' in **35**. You will notice of course that here the cementing of the parts is considerably easier.

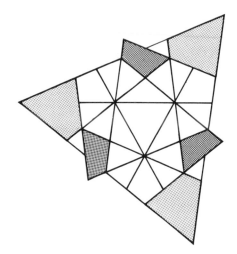

37 Twelfth stellation of the icosahedron

The figure shown here is the net for one part of a platform needed to support the cells whose net was given previously under the caption, 'the core'. Here the core becomes a polyhedral cell, so its vertex must not be removed as was done when it served only as a structural piece. The platform part has wings that belong to the facial planes of **32**. So here the bottom edge has the length of the wedges **29**, but not their sides. The part makes a low triangular pyramid without a base. Five of these pyramids are joined in a ring to make the platform. Then if the tabs at the pentahedral lower edges are turned up they will serve as a groove. The cells may now be placed on the platform, but again it is advisable to cut away the vertex of the pyramid as shown in the figure by the arc mark, because at these points the cells will have their vertices coming in contact and thus at these points the cementing is done. Once a ring of cells has been done it must be allowed to set up rigidly. Then it can be moved and again each ring of cells will be completed in a similar fashion by moving the model in various stages of completion so that the new cells always lie on the platform. A dodecahedral shell will again make a good support for stabilizing the model as it is given its various positions, but it may also be dispensed with. This means if you use it that the platform may be placed in the bottom of a dodecahedral cradle with one pentagon removed to make it possible for you to work on the parts. This model is strikingly pleasing in its simplicity and openness.

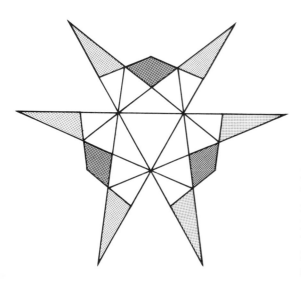

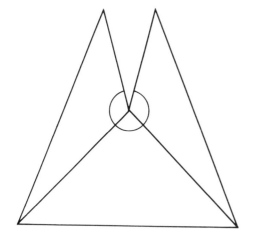

38 Thirteenth stellation of the icosahedron

The net shown here can be used to make a beautiful polyhedron with twelve long penta-hedral spikes each of which is surrounded at its base by five shorter trihedral spikes. You will easily recognize both of these, as they have appeared in previous models. The icosahedral arrangement of colours will give you the longer spikes with facial planes the same colour, but the shorter spikes will be like those described for **35**. The technique for assembly here is to do the short spike first. Cut the card so no tab remains on the right side of the long part and continue this cut to the first interior line as shown. The two triangles on either side may now be folded down and their remaining edges brought into contact to form the short spike. Finally the small triangle at the bottom is folded down so the two acute vertices coincide. Now five of these parts are joined to make the long spike. This will turn out to be a complete poly-hedron in itself since the bottom of the spike becomes entirely enclosed. But now the five parts whose icosahedral colour arrangements are (1), (2), (3), (4), (5) can be laid out on the same star-patterned pallet used for **30**. The obtuse vertices around the base of the spikes will then be found to be coinciding, one blunt vertex of one spike touching one on an adjacent spike. Theoretically the small spikes should also have acute vertices in contact, but this is almost im-possible to achieve in practice on all the small spikes at once. So it is advisable to cement the obtuse vertices forming a ring of parts and as many small spikes as can conveniently be done. The rest of the procedure is the same as for **30**. Once the whole model is finished you can gently apply pressure to the smaller spikes and adjust them, as needed.

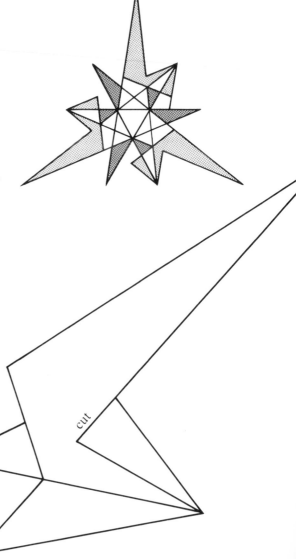

39 Fourteenth stellation of the icosahedron

The net shown here leads to an attractively twisted polyhedron. It has some of the usual dodecahedral symmetry, like the deltahedron **28**, but it also has pentagonal holes leaving the interior hollow and open to view. The method of assembly is to follow the usual icosahedral colour arrangement, the same as in the deltahedron **28**. You may begin with the (0) section. This will be the usual dimple but with a hole in the middle. The tabs are shown on the net to indicate how one edge is broken; the upper tab of this edge serves to join a ring of five parts forming the dimple, and the lower tab serves to join edges forming a trihedral angle on the interior. One of these trihedral angles is not completely formed until three sections are assembled. You will need a bit of skill to reach the interior as the model progresses, but you can apply the cement with a probing needle, working it in between the tabs, and then applying pressure with your fingers along the interior edges till the cement is set. This model calls for patience and accurate workmanship, since the assembly is very intricate. But it makes a very interesting model.

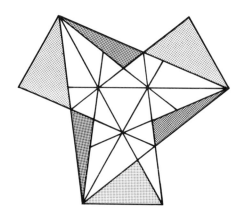

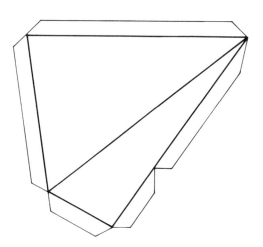

40 Fifteenth stellation of the icosahedron

You can use the net shown here to make another polyhedron with the same attractively twisted appearance as the previous one. This one has the sixty short trihedral spikes arranged so that the interior of the polyhedron remains hollow and is visible only through narrow chinks. You may begin by making five parts as shown. The tabs are indicated and a special cut as well; the other lines are folds in the usual manner. This may now be formed into one of the short spikes. When you have five of these ready they can be assembled to form the usual pentahedral dimple from which the spikes radiate. The cut will receive a tab which has been turned outward from a neighbouring part and the parts may thus be cemented with the help of some pressure exerted by the fingers or with tweezers from the outside. This completes one section. Twelve sections are needed. They are assembled with the aid of the same deltahedral platform used before. A dodecahedral shell may be used so that the interior of the shell provides side supports as the sections are being cemented. The cementing is done at the obtuse vertices which come in contact. It is not always possible in practice to cement all the sharp vertices of the small spikes at once. But as the model is completed some adjustments become possible after the cement is set. Careful workmanship will give good results.

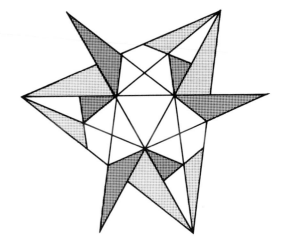

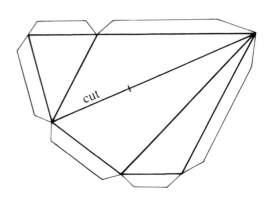

cut

$$\tfrac{5}{2}|2\ 3 = \{3, \tfrac{5}{2}\}$$
$$20\{3\}$$
$$5^{\frac14}\,\tau^{-\frac12}$$

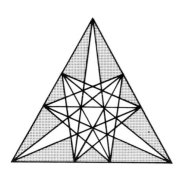

41 Great icosahedron

Of all the polyhedra so far described perhaps the most beautiful and attractive is the great icosahedron, the last of the four regular (Kepler–Poinsot) star polyhedra. The vertex figure here is a regular pentagram. In this respect it is like the great dodecahedron. These two solids stand alone as the only regular star-vertexed polyhedra. You will see many with star faces among the uniform polyhedra to be presented later in this book, but none of them is star vertexed. A model of the great icosahedron is not hard to make. The nets are simple and when the model is assembled as described here it is also very sturdy and rigid although it is completely hollow inside. Doing it in five colours takes a little more time, but it is well worth the effort. The paired arrangement of parts and the colour table are given below.

A set of five pairs makes the fan-like form shown. You must now see to it that the folding is done so that it is down between each member of the pairs and up between the pairs, accordion fashion. Then by cementing the remaining edges a vertex part for the great icosahedron begins to

	12	34	56	78	9X
(0)	YG	BY	OB	RO	GR
(1)	BG	YB	RY	OR	GO
(2)	OY	BO	GB	RB	YR
(3)	RB	OR	YO	GY	BG
(4)	GO	RG	BR	YB	OY
(5)	YR	GY	OG	BO	RB

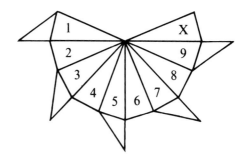

take shape. The smaller isosceles triangles are then cemented in place to form a pentagonal dimple from which the vertex part rises. Twelve of these are needed, the first set of six as set out in the colour table and the second set of six in the enantiomorphous order. These vertex parts are then joined following the icosahedral arrangement, enantiomorphous pairs being diametrically opposite each other. In cementing the vertex parts together, give your attention to only one edge at a time. Clamps can readily be used because the dihedral angle between two adjacent facial planes along the pentagonal edges of a vertex part is very acute. Even the last vertex part goes on in the same way. On the last edges the cement is applied carefully to the crack and then worked down between the tabs with the probing needle before setting the clamps. If you find that there are openings or small holes left at the corners of the pentagonal edges, do not be dismayed. You can close these after the model is completed by adding a drop of cement to each hole, working it in with the probing needle and then applying a little pressure when the cement begins to harden. You will find that this closes the holes successfully and at the same time adds rigidity to the model. This model, when it is well made, is always a delight to behold.

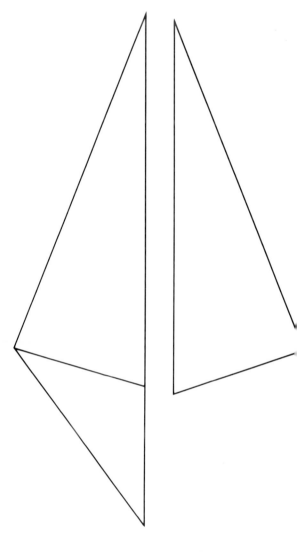

42 **Final stellation of the icosahedron**

This is the final stellation of the icosahedron. It is a spiny-looking polyhedron, the spines falling into fairly well-defined clusters of five. Twelve of these clusters complete the solid. A net for making a model of this polyhedron is shown below. It will not give the facial planes each its own colour, but then the work is reduced by using the net as shown. Sixty of these trihedral spikes are needed. A set of five joined in a ring makes one section or the cluster mentioned above with the bottom forming the edges of a pentagon. So twelve of these are joined in the usual dodecahedral fashion. It is best here to reduce the scale of the model, unless you want to show its actual size in relation to the great icosahedron. If the latter is about 10 inches high, the final stellation on the same scale is nearly 24 inches high. If the model is done as suggested here, completely hollow inside, the smaller scale gives a better, more stable, result. The sixty spikes radiating from the central mass of this polyhedron make it look like rays of light emanating from the sun.

If you are interested in making more of these stellated icosahedra you can work out your own nets by consulting *The fifty-nine icosahedra*, by Coxeter, Flather, Du Val and Petrie. In this booklet all the cases are illustrated with excellent drawings of both the solids and their facial planes. Each one presents its own challenge and will give you a sense of satisfaction upon completion.

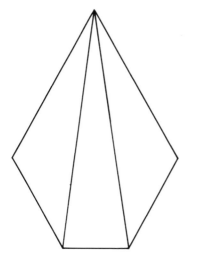

Commentary on the stellation of the Archimedean solids

In the previous pages you saw the stellation process applied to the Platonic solids. You may now be wondering whether the Archimedean solids can also be stellated. The answer is that they can. The procedure is the same: each facial plane must be extended indefinitely to generate cells exterior to the original solid. Using these cells as building blocks, you can form many new solids, theoretically at least. In practice however, for the purpose of making models, the stellation pattern is more useful, although you will find it helpful to have some acquaintance with the cells as well. Nets for these cells can easily be found from the stellation pattern. But you may be asking yourself: why should anyone want to stellate the Archimedean solids? Isn't it a lot of work? Yes, the work involved here does begin to look overwhelming. It really calls for team work. It is hard to find much published material on this topic, and undoubtedly it is because of the great number of possible forms that come crowding up for consideration. A complete enumeration of all the possible stellations is a mathematical question that has yet to be investigated. No doubt some restrictive rules, such as J. C. P. Miller designed for the case of the icosahedron, would have to be applied here also. These forms do not always turn out to be particularly attractive or aesthetically pleasing, but then again many of them do. The final stellation usually seems to be of more than ordinary interest.

A more important question mathematically is: Are any of these stellated Archimedeans regular or uniform polyhedra? Before attempting any answer to this problem, you will find it enlightening to see the stellation process applied to at least two Archimedeans.

The cuboctahedron and the icosidodecahedron are given here because of their close relationship to the dual Platonic pairs. Also as quasi-regular solids they should prove to be the most interesting or the most likely ones to generate further regular or uniform polyhedra.

How is the stellation pattern arrived at? If you look back at the case of the octahedron, you will notice that the pattern is actually a set of six lines. They can easily be counted as they come in parallel pairs (see fig. 28). The inner triangle is one face of the original octahedron. If you set a model of the octahedron on the drawing so that one of its faces exactly covers this inner triangle, the other lines are easily seen to be the intersections of the other facial planes with the plane of this base triangle. Since the triangle on top is directly opposite the base triangle, it is on a plane parallel to the plane of the paper and so it generates no line on the stellation pattern. Thus the eight faces of the octahedron are all accounted for.

If you turn now to the dodecahedron, its twelve faces ought to give a stellation pattern of ten lines. This is indeed the case; five parallel pairs appear (see fig. 29). If you set a model of the dodecahedron on this drawing, so that one of its faces coincides exactly with the innermost

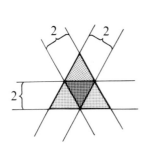

Fig. 28

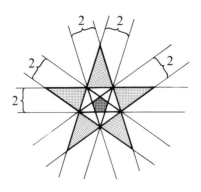

Fig. 29

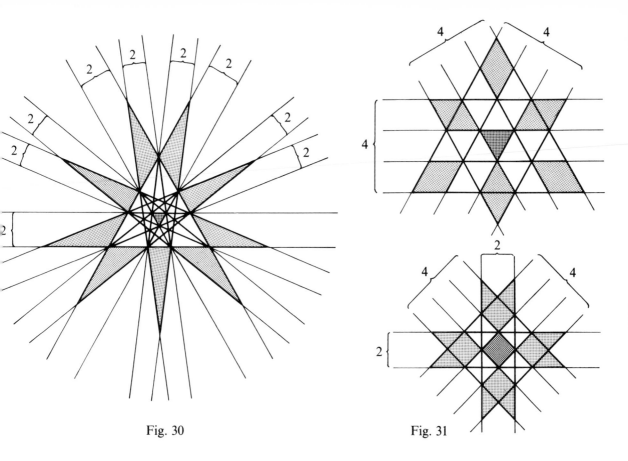

Fig. 30

Fig. 31

pentagon, you can move your eye into the plane of the other faces produced and observe how they cut the plane of the paper precisely on the other lines of the drawing.

The icosahedron gives a similar result. Its twenty facial planes generate a stellation pattern of eighteen lines, nine parallel pairs (see fig. 30).

You can now see what principle is at work in the stellation process. Applying it to the Archimedean solids simply means that two or more stellation patterns turn up, one for each face that is different.

The cuboctahedron has eight triangular faces and six square faces, a total of fourteen. So it will have two stellation patterns, each with twelve lines. For the triangular face there are three sets of parallel lines, four to a set; for the square face there are four sets of parallel lines, but they come in one set of two, then a set of four, and then a repetition of these, a set of two and a set of four (see fig. 31). The cells are easily enumerated. There are only four different kinds. Besides the cuboctahedron itself you will find

six square pyramids with equilateral triangles for side faces, eight triangular pyramids with triangular bases and isosceles right triangles for side faces, twenty-four dipyramids with equilateral and isosceles right triangles for faces, twenty-four pyramids like the six before, and finally twenty-four rhombic pyramids with rhombic bases and equilateral and isosceles right triangles for side faces. How many solids can be formed from these cells? That all depends on what restrictive rules you wish to apply. Some may be only vertex connected or they may have holes leading to the interior, like some of the stellated icosahedra. Miller's rules for the icosahedron invoke chiefly symmetry requirements and accessibility from the outside. The models **43, 44, 45** and **46** are examples of only four possible cases.

As before, the shaded portions of each facial plane show what is visible from the outside of the polyhedron. It is also from these shaded portions that the nets are derived for the construction of the models.

43 Compound of a cube and octahedron

The first stellation of the cuboctahedron is a compound of the cube and the octahedron. This makes a very interesting model when done in colour. The cube may be done in three colours and the octahedron in two more. The nets are simply the two different types of triangles shown in the facial planes below. Four of the equilateral triangles are cemented with alternate Y and B parts to make a square pyramid without its base. Three of the isosceles right triangles make a triangular pyramid, again without its base. The colours O, R, G are arranged so the opposite facial planes of the cube have the same colour. Six of the first kind of pyramid and eight of the second will complete the model. This hollow model gives a neater result than can be obtained by cementing pyramids of one kind to a basic cube or octahedron. A net with one long tab crossing a weak point, as shown here, can also be used to make a very rigid model.

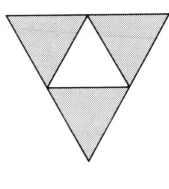

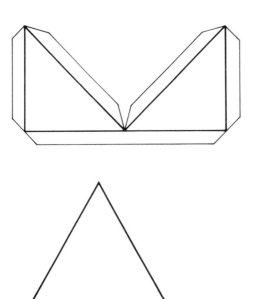

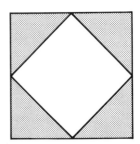

44 Second stellation of the cuboctahedron

This polyhedron arises as the second stellation of the cuboctahedron. Twenty-four dipyramids have here been added to the compound of the cube and the octahedron. A model of this polyhedron, however, is more easily made using the same nets as before. Here it is best to make a ring of four parts using the net with one long tab crossing a weak point. These four parts form a square section prism, the long tabs being at the square bottom and the jagged sides coming right down to the bottom at the centre point. The top of this prism is then closed with eight equilateral triangles. This completes one section of the model. Six of these sections are needed altogether. The colour arrrangements of the facial planes may follow those of the previous model as they extend themselves to this case. The final result is easily recognizable as a truncated form of the stella octangula.

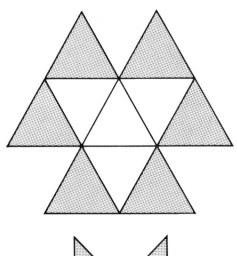

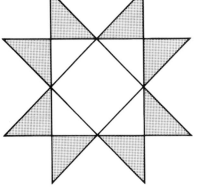

45 Third stellation of the cuboctahedron

This polyhedron is very interesting in more than one way. First of all the square faces stand out very plainly falling as they do into three sets of pairs. The members of each pair are parallel to each other and the sets are perpendicular to each other. Secondly the polyhedron is a sort of compound, composed of six square pyramids, the squares mentioned previously serving as the bases of these pyramids while the triangular side faces quickly disappear into the interior of the solid, their vertices coinciding with the middle point of the opposite depression. All this is more readily seen in a model than it is in words or description.

To construct this model, begin with four squares and cement a pair of equilateral triangles to each of them. These parts are then cemented together by joining the triangles, the squares remaining to the outside of the ring. Six of these sections are required, and again they are joined by the equilateral triangles. Finally, trihedral dimples whose faces are the small isosceles right triangles, close the holes formed by three of the connecting triangles.

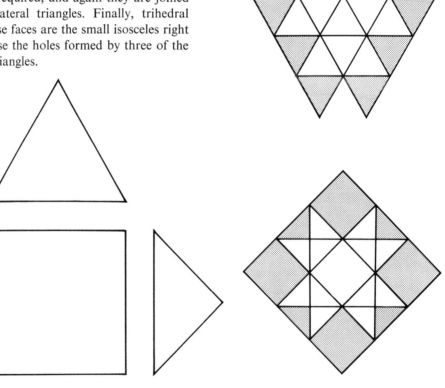

46 Final stellation of the cuboctahedron

The final stellation of the cuboctahedron is particularly attractive since it is a compound of two tetrahedra, Kepler's stella octangula being the final stellation of the octahedron, and of three perpendicular prisms of square cross-section whose common portion on the interior is the original cube. The outer or end faces of these prisms are four rhombic faces forming deep tetrahedral dimples. Two very simple nets serve here for the parts needed to make a model of this polyhedron. The method of assembling in sections is the best to use. Four of the chevron-shaped parts are cemented in a ring to form a prism open at both ends. Four rhombic parts are then cemented to form the deep tetrahedral dimple and this is used to close the upper end of the prism. The cementing is best done one edge at a time and clamps can easily be used. Six of these sections are needed and they are joined by the vertex parts of the stellated octahedron. These vertex parts also have rhombic faces, so the same net serves for them as for the dimples. In doing the last section it is best to cement it in

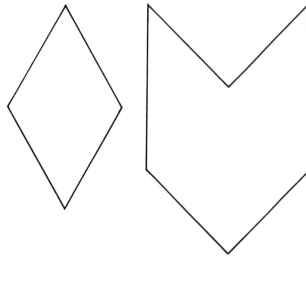

place before closing its end with the rhombic dimple. In this way you can still work on the tabs through the open end of the prism which can easily be closed last of all.

You can see that none of these stellated cuboctahedra is regular or uniform. But maybe some others not shown here would turn out to be so. To find them you would have to investigate the stellation pattern and find regular polygons whose line segments coincide with the lines of the pattern. The first model given, **43**, fulfills this requirement. However a second requirement is that the result be a genuinely new polyhedron, not a compound. You can see that this is not fulfilled. The first model, **43**, is a compound of a cube and an octahedron. Model **44** has one facial plane regular, the octagram, but it is combined with a truncated triangle which is not a regular polygon; so that fails. The third one, model **45**, also comes out with one face regular, the square, but it again is combined with a truncated triangle. The final stellation has neither of its faces regular. But it does bear some resemblance to **92** of the uniform polyhedra.

Commentary on the icosidodecahedron

The icosidodecahedron has twelve pentagons and twenty triangles as faces, a total of thirty-two. This looks a formidable number to investigate. It will mean studying two stellation patterns, each composed of thirty lines.

As a first step to drawing the patterns it is worth noting that an icosidodecahedron is the solid that is common to a compound of a dodecahedron and an icosahedron. These patterns are already known. Those to be done now must therefore bear some resemblance and ought to serve as guides. So proceeding as for the stellated cuboctahedra you should be able to verify the two patterns given on the following pages.

These two stellation patterns lead to forty different kinds of cells. You would have to be exceptionally ambitious to verify this, so no more will be said about them here, except for this one brief remark. Just as the icosidodecahedral patterns include as a subset the lines of the dodecahedral and icosahedral patterns, so too the icosidodecahedral cells are building blocks for the dodecahedral and icosahedral cells. In other words these latter cells are further split up and subdivided by the extended facial planes of the icosidodecahedron.

The following polyhedra exhibit only a representative sub-set of stellated forms in this set. You will quickly notice that some of them are compounds or variations of the three stellated dodecahedra and/or some stellated icosahedron. This gives many of them the same beauty of form. But—the big question—are any uniform polyhedra to be found? The answer seems to be—no. Striking resemblances turn up, but none of them satisfies the definitions of a uniform polyhedron.

As for finding regular polyhedra it has already been mentioned that the great mathematician Cauchy in 1811 proved that the four Kepler–Poinsot solids taken together with the five Platonic solids exhaust the list of regular polyhedra. So if you continue to look, you are merely joining the ranks of angle trisectors or circle squarers or cube duplicators.

For identification the models are numbered **47** to **66**.

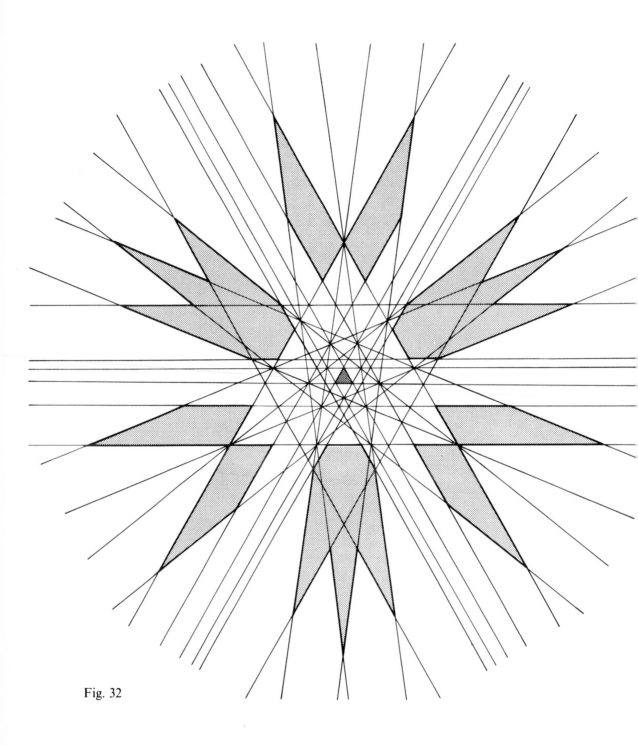

Fig. 32

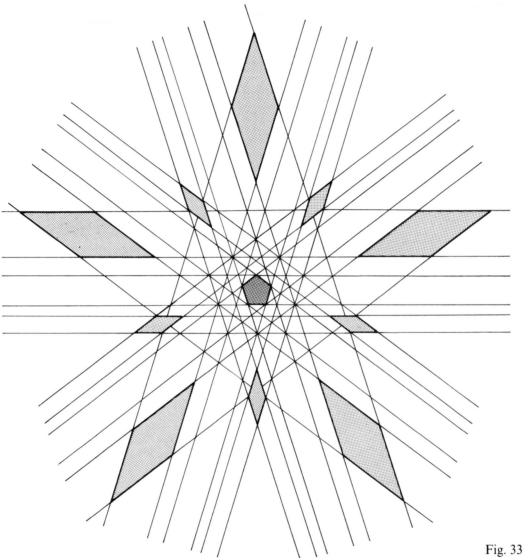

Fig. 33

47 First stellation of the icosidodecahedron

This polyhedron is a compound of two Platonic solids, the dodecahedron and the icosahedron. It is the first stellation of the icosidodecahedron. This begins what is called the main line, namely the polyhedra derived successively from previous ones by adding cells to cover completely all the 'outside' surface area. Thus twelve low pentagonal pyramids and twenty small triangular pyramids cover completely the interior icosidodecahedron. A model, however, can be made completely hollow inside. You may make the pentagonal pyramids following the icosahedral arrangement of colours, but it is best to make all the small triangular pyramids the same colour, say W. These pyramids, that is both types, are left without bases. They are then cemented to one another in the same way that you cemented the parts for any convex polyhedron. This method of construction gives a very neat result.

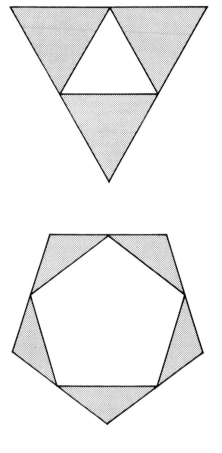

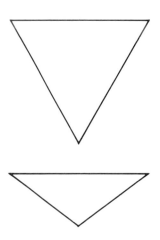

48 Second stellation of the icosidodecahedron

This is the second stellation of the icosidodecahedron in the main line. It is related to the small stellated dodecahedron, being in fact a special type of truncation of that regular star polyhedron. This suggests a method of construction.

A net of five parts is shown with the tabs illustrated also. The triangles may be arranged in the icosahedral colour scheme, ten of them being required to close the opening once the ring of five parts has been formed. Then these sections are cemented as usual in the case of a dodecahedron. The tabs give it good rigidity since they span a crucial weak point. This polyhedron is not particularly attractive—merely a stepping stone to other forms.

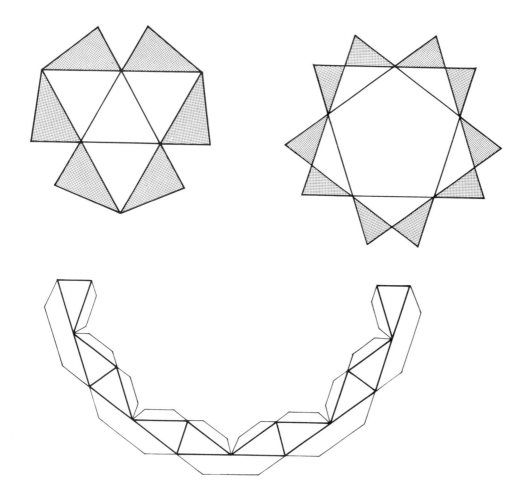

49 Third stellation of the icosidodecahedron

This is the third stellation in the main line of the icosidodecahedron. It is interesting to see a pentagon as one face, but the other is not a regular polygon. In making a model of this polyhedron it is best to make sets of trihedral parts which are vertices of the polyhedron. These are formed by using one rhombus and two triangles for each part, as shown. Five of these form a ring. The rings are cemented together using the small dimples and triangles as one connector between three rings. This may not be a very attractive model, but it helps to illustrate the cell formation of polyhedra. You may follow the colour arrangement through from the first two models, making all the pentagon parts W and using the five-colour icosahedral arrangement for the rest.

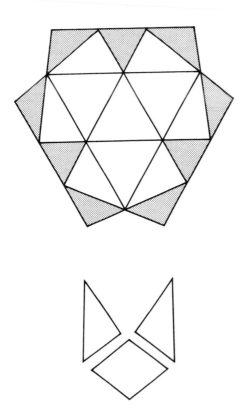

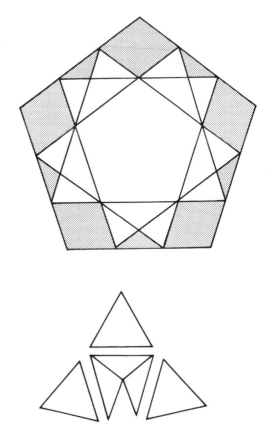

50 Fourth stellation of the icosidodecahedron

This polyhedron is a compound of the small stellated dodecahedron and the triakis icosahedron, both the first stellations of the dodecahedron and the icosahedron respectively. The facial planes shown above make this relationship evident. An attractive model results from following the usual colour arrangement used in the original separate polyhedra. Thus you may make the vertex parts of the small stellated dodecahedron in the form of pentahedral angles and the vertex parts of the triakis icosahedron as trihedral mounds whose faces are three kites. These mounds then become the connectors for the other vertex parts. The photograph will make the arrangement evident.

This polyhedron is no longer in the main line of stellation. It is more interesting to follow out various combinations of compounds. Since the dodecahedron has three stellated forms and the icosahedron has fifty or more to choose from, the results can become something like musical 'variations on a theme'.

51 **Fifth stellation of the icosidodecahedron**

In this polyhedron the small stellated dodecahedron penetrates the compound of five octahedra. To construct a model you may proceed as before. First make the pentahedral vertex parts of the small stellated dodecahedron in the usual colour arrangement. But notice that the shape is slightly different toward the bottom. The net for the vertex parts of the compound of five octahedra is simply a set of four quadrilaterals derived from the appropriate facial plane shown above. These again form mounds, slightly higher than in the previous model. These mounds again serve as connectors for the pentahedral vertex parts. However they are also connected to each other in three's. This will be evident to you from the photograph. The octahedra here can easily escape notice. They only become apparent when the model is examined at close range.

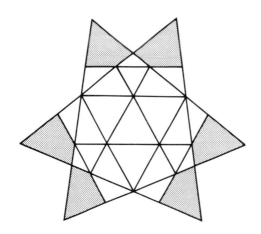

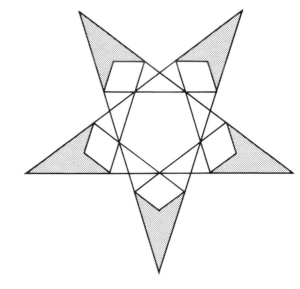

52 Sixth stellation of the icosidodecahedron

This polyhedron has two pentagrams coinciding with the twelve facial planes of a dodecahedron, one of the pentagrams (the larger one) still belonging to the small stellated dodecahedron. It can be derived from the previous polyhedron by the removal of two different types of cells, which alter the shape of the compound of five octahedra to the resulting form it assumes here. The same technique of construction that was used in the two previous models can be used here. The twelve vertex parts of the small stellated dodecahedron are done in the usual way and a hexahedral mound serves as a connector. The faces of this mound are the following: a kite from the smaller pentagram, then two triangles which are enantiomorphous pairs from the icosahedral plane, then another repetition of these—a kite and a pair of triangles. The drawings and the photograph together should make this clear.

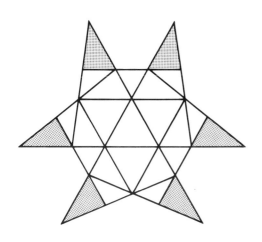

53 Seventh stellation of the icosidodecahedron

This polyhedron is a compound of the great dodecahedron, the second stellation of the dodecahedron, and **32**, one of the stellations of the icosahedron. To build a model of this polyhedron the best procedure is to make the dimples of the great dodecahedron with a large hexagonal hole in each, following the usual colour arrangement. Next prepare the hexahedral spikes, whose faces come from the icosahedral planes, according to their colour arrangement. These spikes are then cemented to the dimples, closing the holes. These sections are then assembled as for the great dodecahedron. The single tab crossing the weak point gives the rigidity needed for a beautiful model.

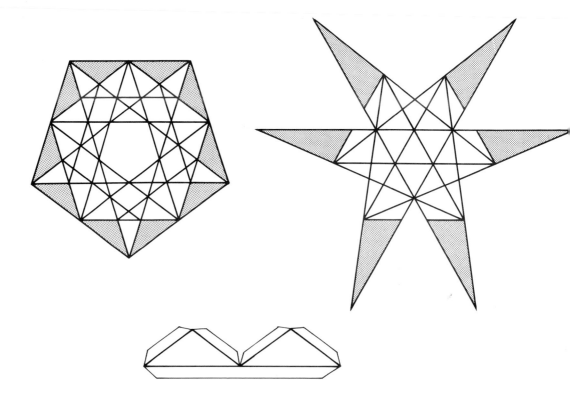

54 Eighth stellation of the icosidodecahedron

This polyhedron is easy to recognize as the compound of five tetrahedra penetrated by the great dodecahedron. The vertices of the latter appear as small rosettes at the bottom of the dimples of the former. If you were successful in making a model of the compound of five tetrahedra, you will undoubtedly want to attempt this one also. The procedure in making this model is practically the same. You may begin with the trihedral vertices of the compound of five tetrahedra and join them in a ring, as before. In this case, of course, a decagonal hole is left in the centre of the ring. Fill this hole with a vertex part of the great dodecahedron, the rosette mentioned above. It is easy to see, from the pentagon plane, that this has ten triangular faces. The secret in cementing these rosettes is to do them one tab at a time. You will need considerable skill and patience toward the end of the work. But it is worth the effort, because it makes a very beautiful model. The colour arrangements are the usual ones used in the separate polyhedra.

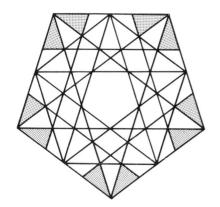

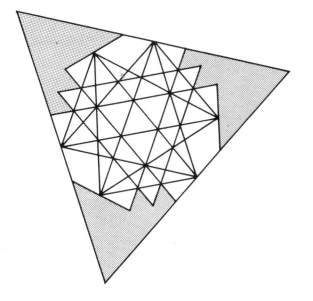

55 Ninth stellation of the icosidodecahedron

This polyhedron is the compound of ten tetrahedra, but the ghost of a great dodecahedron leaves the traces of its facial planes in the holes at the bottom of the dimples and on the interior which remains visible through the holes. The rosettes used in the previous case would exactly fill these holes, but this makes a very interesting model as it is. However, it calls for a slightly different method of assembly. Two of the butterfly shapes may be joined with two pairs of the smaller triangles, the grooves between them acting as connectors. The V-cut at the bottom, that is, under the groove, then has two of the irregular pentagons as faces. These faces eventually form the interior surfaces. This assembly forms one section. Thirty of these sections will complete the model.

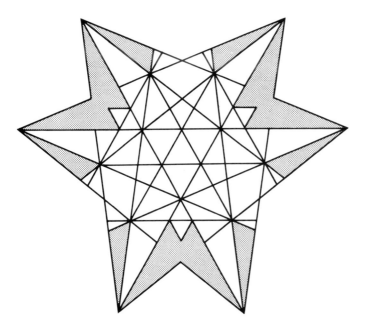

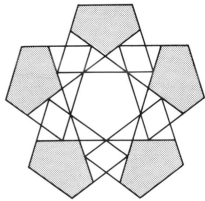

56 Tenth stellation of the icosidodecahedron

This simple polyhedron has the external appearance of the deltahedron **28**, one of the stellations of the icosahedron. Here the ghost of a small stellated dodecahedron leaves its traces in the hole at the bottom of the dimple and on the interior. The vertex parts of the small stellated dodecahedron would fill these holes. This means that the same thing could be done to the compound of five tetrahedra and the compound of ten tetrahedra; the adjustments to be made to the nets are easily seen from the stellation patterns.

The method of assembly for this polyhedron had best be like that used in the previous model. One section has two truncated triangles joined along the longest side and two isosceles trapezia which eventually become the interior of the model. These sections are rather flexible but once they are joined rigidity is achieved. Thirty sections complete the model.

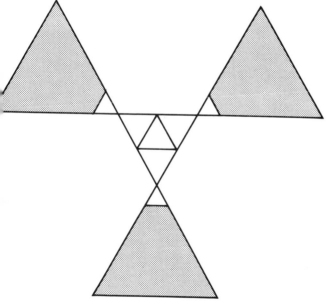

57 Eleventh stellation of the icosidodecahedron

This polyhedron is truly remarkable because it bears such a close resemblance to one of the uniform polyhedra. As the drawings here show it has one facial plane an equilateral triangle and the other a decagram, almost regular but not quite. It is actually a truncated form of the great stellated dodecahedron, the truncation occurring very low, near the base of a vertex part. This fact suggests a method of assembly for a model. You may join three of the parts shown below. Keeping one tab long strengthens the weak point. Three kites form a dimple and are used to close the opening of the truncated pyramid part. These parts are then cemented in the same way as was done in the great stellated dodecahedron. The usual dodecahedral and icosahedral colour arrangements can be used very effectively.

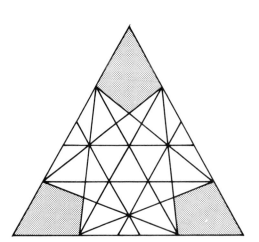

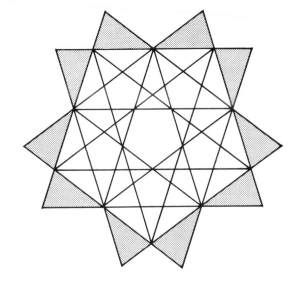

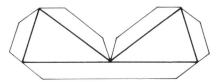

58 Twelfth stellation of the icosidodecahedron

This polyhedron is very attractive. A truncated form of the great stellated dodecahedron is here making its appearance, penetrating a delta-hedron, like the one found among the stellated forms of the icosahedron, **28**. The icosahedral faces are composed of three equilateral triangles, slightly larger than those in **28**, so that their common portion turns out to be the triangular faces of the interior icosidodecahedron. The dodecahedral faces are decagrams, almost regular in appearance but not quite so. It is the icosahedral faces which have truncated the great stellated dodecahedron. As you will see once you have constructed a model of this polyhedron, it also gives the appearance of five long cells tapered at their outer ends all placed neatly in a ring in each dimple of the deltahedron, **28**. This fact suggests a very simple way to construct a model of this solid. Begin by making a set of five long cells with the four parts as shown (note that they are not closed cells). These cells are then cemented in a ring, adjoining each other and radiating outward from their blunt ends. Portions of the triangular planes are cemented between these cells, giving a section reminiscent of those used in other stellations of the icosa-hedron, in particular **28** and the great icosa-hedron **41**. Twelve of these sections will complete the model.

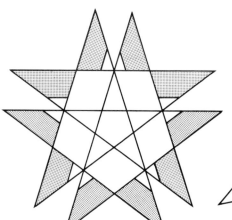

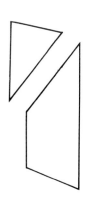

Portion of triangular plane

The four parts for a long cell (not closed)

In the drawings given on the previous page only some of the lines from the stellation patterns are shown. You can easily see for yourself how this is so. You will then also notice how the vertices of the great stellated dodecahedron just manage to make it to the exterior lines of the dodecahedral pattern. This also means that the vertex parts of the solid are dissected into numerous cells. Various selections of these cells could lead to many different truncated forms of the great stellated dodecahedron.

59 Thirteenth stellation of the icosidodecahedron

In this polyhedron the great stellated dodecahedron penetrates **34**, one of the stellations of the icosahedron. Twenty plus twelve vertex parts are thus seen radiating from the central mass. If you have succeeded in making models of the two separate polyhedra, this compound should not give you too much trouble. The vertex parts of both types are first assembled separately and then joined to one another. You may use the usual colour arrangements and begin by surrounding an icosahedral vertex part with a ring of five dodecahedral vertex parts. Once this is completed the rest becomes evident. The secret in doing the last part is this: cement the three kites of a dodecahedral vertex part separately, not as one unit. In this way the three longer sides of these kites are cemented last of all and can be pinched with the fingers from the outside. This makes a very attractive model.

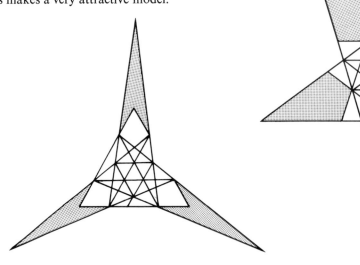

60 Fourteenth stellation of the icosidodecahedron

In this polyhedron the great stellated dodecahedron penetrates a truncated form of the great icosahedron. The truncation is effected by removal of some of the stellation cells from the latter. It certainly gives a spiny result resembling a sea urchin. The twelve vertex parts of **34** are still seen here surrounded by a ring of five lower vertex parts. The vertex parts of the great stellated dodecahedron exhibit rhombic faces.

A good method for constructing the model is to assemble the three parts needed for the lower vertex part, then to assemble a ring of five. This ring will have a pentagonal hole at the bottom of a depression, its edges formed by the small isosceles triangle with base angles 72°. This pentagonal hole is closed with the long spiked vertex part of the stellated icosahedron **34**. This is one section of the model. Twelve of these are needed and they are joined together using the dodecahedral vertex parts as connectors. The last part may be completed as in the previous model.

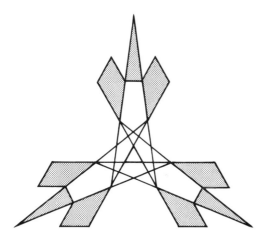

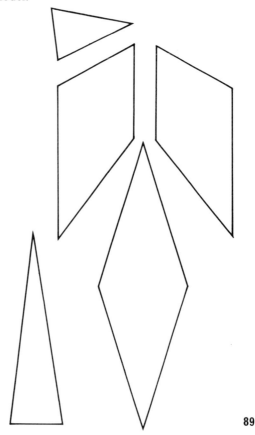

61 Compound of the great stellated dodecahedron and the great icosahedron

The two regular star polyhedra, the great stellated dodecahedron and the great icosahedron are probably the most attractive polyhedra of all. Here they are to be found together in one compound, different from the one given by Cundy and Rollett, *Mathematical models*, pp. 132–3. Here they appear together as a stellated form of the icosidodecahedron. To make a model of this polyhedron you may use the same technique as that used for the great icosahedron. The nets are shown below. These vertex parts are not as stable as the former ones, since some of the lower portions are missing. But these get their rigidity back again when the dodecahedral vertex parts are added in rings of five. You will find that it will take a great deal of patience to do this model well. But then it is also well worth the effort.

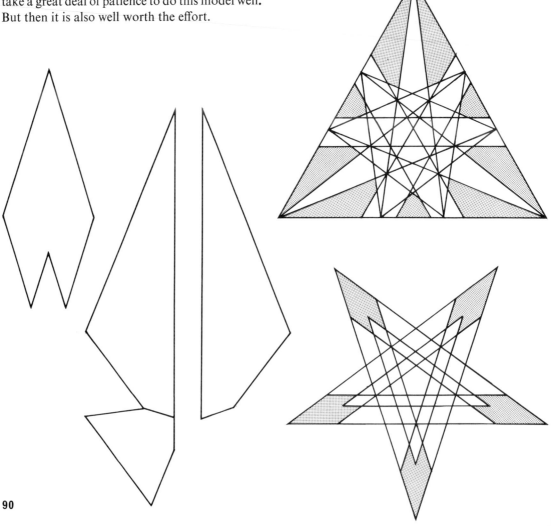

62 Fifteenth stellation of the icosidodecahedron

This is a remarkable polyhedron because it so closely resembles number **95** among the uniform polyhedra given later in this book. It is not itself uniform because the hexagons in its facial planes are not regular and the pentagons are incomplete or broken at the vertices. The drawings of the facial planes shown below reveal this clearly. But it is not difficult to see what adjustments are required to achieve uniformity. This polyhedron is a truncated version of the great icosahedron. The pentagonal planes of the icosidodecahedron are here effecting the truncation. Number **95** is also a truncated great icosahedron, but the truncation is more simply effected by a plane parallel to a plane of the vertex figure in such a way that the triangles are converted to regular hexagons.

A model of this polyhedron can be assembled the same way as that used in the great icosahedron. The nets for the icosahedral planes are shown below; they are merely truncated versions of the other nets. After you have assembled a ring of these a star-shaped hole is left. This is closed with a very elaborately pitted star form. Its centre is a vertex part of the small stellated dodecahedron turned inside out to form a cup. A set of five trihedral dimples forming the star arms are cemented to the edges of the cup. This completes the pitted star. It is cemented into the icosahedral part one edge at a time with the aid of clamps. Twelve icosahedral parts complete the model.

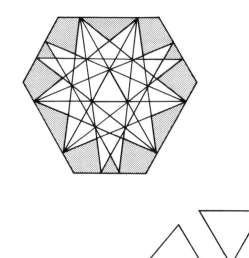

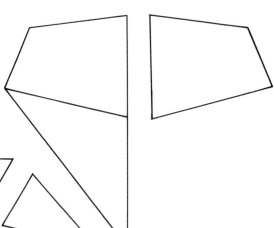

63 Sixteenth stellation of the icosidodecahedron

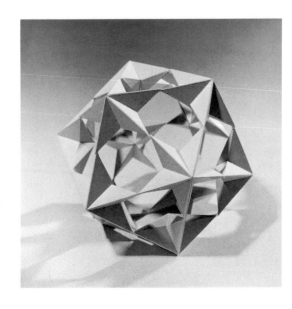

Some of the stellated icosahedra exhibit cells or combinations of cells which are connected only by the vertices. This gives these polyhedra an open, airy quality. The stellation cells of the icosidodecahedron can be selected in the same way, giving the same result. Three of these are presented here, this model and the following two. You may imagine this one as resulting from the removal of the stellated icosahedron **32** from the compound polyhedron **53** (see pp. 53 and 82). The vertex parts of the great dodecahedron are left, all alone in the form of beautiful solid stars, a set of twelve, joined by the vertices. The nets for one star arm are shown below. Five of these pairs are needed for one solid star. To join them at their vertices you will need a construction cradle. This cradle is one section of the great dodecahedron, containing one complete pentagonal plane, inverted of course so that it becomes a cradle. It is best to cut holes in the edges of the cradle at the points where the cementing is done. Patience and a steady hand will give you the beautiful model shown in the photograph.

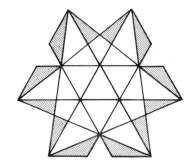

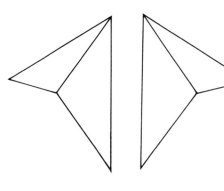

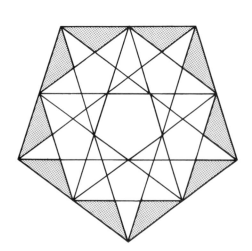

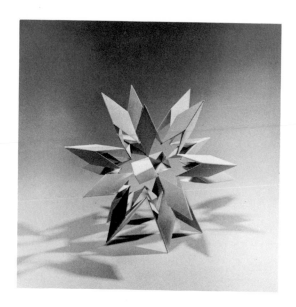

64 Seventeenth stellation of the icosidodecahedron

This polyhedron and the next one are closely related to the compound of the great stellated dodecahedron and the great icosahedron **61**. You may imagine in this case that the great icosahedron has disappeared leaving the traces of its facial planes on the interior of this one. But to get the remaining cells of the great stellated dodecahedron to be vertex connected the rhombic faces of these cells must be completed. This is evident from the drawings below when you compare them with the drawings of the compound. Completing the faces means adding cells to bring the vertices into contact. You can make a model of this polyhedron by using the net shown below. Three of these make one vertex part for the great stellated dodecahedron, a completely enclosed cell trifurcated at its base. Here again a construction cradle is needed to join the cells. This is simply one ring of five vertex parts of the regular great stellated dodecahedron. Again it is best to cut holes in the edges at the points where the cementing is done. You must also leave three face triangles of the cradle uncemented, otherwise the model cannot be removed and turned in the cradle as the work proceeds. The completed model holds together surprisingly well, a little springy but well able to support its own weight, much like the models **35** or **29**.

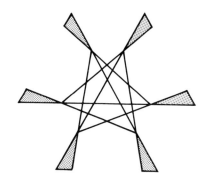

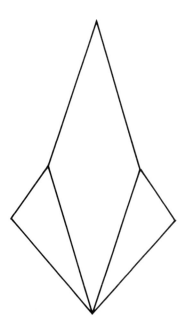

65 Eighteenth stellation of the icosidodecahedron

In this polyhedron the ghost of a great stellated dodecahedron leaves its traces in the holes and in the interior of the great icosahedron. The decagram, really the truncated pentagram, which appears so clearly in the drawing below is virtually lost in the polyhedron because the eye is arrested by the exterior triangular planes. Only close inspection will reveal the true nature of the interior surfaces. A model of this polyhedron is easily made in the same way that you assembled one of the stellated icosahedra, a vertex-connected model like this one. In fact if you make it the same size as **30** you can lay out the cells on the same pallet. A net for one part of such a cell is shown here. Ten of these are needed, five enantiomorphous pairs, for one vertex part. The same colour arrangement as that used for the great icosahedron serves very well here. As you can easily see, twelve of these cells or vertex parts will complete the model. You will also see that the contact points are different here. This makes the model slightly more difficult to construct, but it still comes out quite rigid, surprisingly enough.

66 Final stellation of the icosidodecahedron

The stellation patterns for this model are given on pp. 74 and 75.

The final stellation of any polyhedron is usually of more than ordinary interest. Here is the final stellation of the icosidodecahedron. It gives the appearance of twelve bursting sprays, like fireworks in a night sky, emanating from a central mass, but here all the sprays have mathematical precision. The final stellation of both the dodecahedron and the icosahedron are evident in this polyhedron. The great stellated dodecahedron just manages to reach the exterior as small trihedral cells, almost lost, like blades of grass at the foot of giant oaks. The final stellation of the icosahedron is itself a set of twelve clusters of five long spikes to a cluster. Here five thin cells fill some of the space between these spikes, giving the whole section the spray-like effect mentioned above. This suggests a method for constructing a model of this polyhedron.

Begin by making a cup or tapered prism, open at both ends, one end very jagged and the other

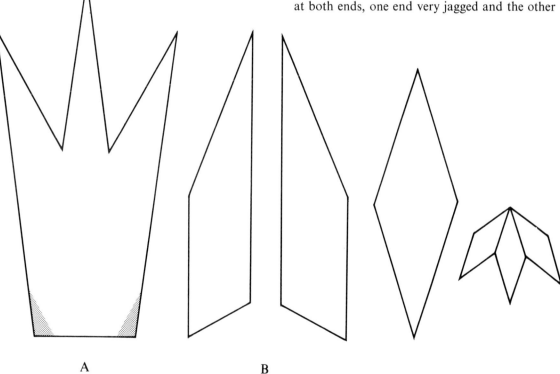

A   B

in the form of a regular pentagon. Five of the parts marked A are used for this. Then five pairs of the parts marked B, ten trapezia, are cemented together, being joined at the lower, short, blunt ends. These parts fold up, accordion fashion, enantiomorphous pairs facing each other. Next cement five rhombic parts so their lower acute vertices go between the openings of the accordion folds. Once this is done you have the deeply pitted interior of a cup or tapered prism. This can now be cemented into the cup, one edge at a time. It is best to do the rhombic edges first since clamps can easily hold at these edges, the dihedral angles here being very acute. The other edges along the centre spike on each of the five sides of the cup can then be cemented, and held in place with your fingers until the cement is set. This completes one section. Twelve sections are needed and these are joined together in dodeca-hedral fashion. Finally the small vertex parts of the great stellated dodecahedron are cemented in place after all the rest is complete. It is easiest to leave all the tabs on these parts and simply apply cement to all six tabs of one part and press it in place between three of the cups. The shaded portion of the net A shows the area that is covered by these parts. This model calls for careful workmanship. The final result can be very attractive in the usual icosahedral colour arrangement with W for all the dodecahedral planes.

You should now be able to discover other stel-lated forms of the cuboctahedron or the icosi-dodecahedron by yourself. This can be done if you are acquainted with the stellation cells and see how their faces are found in the stellation patterns. In fact you should now also be able to stellate other Archimedean solids by yourself. You do not need complete stellation patterns to begin. These are put together as you proceed, by trial and error if by no other way, something like a crossword puzzle or, even more so, like a three-dimensional jig-saw puzzle.

III Non-convex Uniform Polyhedra

Commentary

You have now seen the stellation process applied to the Platonic solids and two of the Archimedean solids. You have also seen that it leads to very few uniform polyhedra. In fact only the three dodecahedral stellations and one icosahedral stellation turn out to be uniform. You may recall that a polyhedron is uniform if all its faces are regular polygons (these now include the regular stars) and all its vertices are alike. The list so far contains the five Platonic solids, the thirteen Archimedean solids and the four Kepler–Poinsot solids. Are there any more uniform polyhedra? It may surprise you to learn that there are at least fifty-three more! How were they ever discovered? Thirty-seven of them are due to Badoureau (1881) who systematically considered each of the Platonic and Archimedean solids in turn with a view to finding regular polygons or regular stars on their facial planes or cutting through the interior of these solids. This is a different approach from that of stellation. If such a polygon is found, it is evident that its vertices coincide with some of the vertices of the related convex polyhedron. The planes of these polygons may intersect. If portions of the solid are removed symmetrically, another uniform polyhedron may result. This process is called faceting, a sort of reverse of stellating. Stellating implies the addition of cells to a basic polyhedron which serves as a core. Faceting implies the removal of cells, so that the basic polyhedron may still be imagined as a case or enclosing web for the new one. If you examine Kepler–Poinsot solids from this point of view you will see that the small stellated dodecahedron and the great dodecahedron can both be derived by faceting an icosahedron. The vertices of the former and the edges of the latter coincide respectively with the vertices and edges of an icosahedron imagined as a case enclosing them. If you examine the models you will see this very

plainly. The great stellated dodecahedron is a faceted dodecahedron as well as a stellated one. If you imagine straight lines joining each vertex to three adjacent ones, the whole set of these line segments forms the edges of a regular dodecahedron. Thus the vertices of the great stellated dodecahedron coincide with those of a dodecahedron encasing it. The great icosahedron is a faceted icosahedron as well as a stellated one, for the same reason. Many of the models now to be presented will amply illustrate this principle of faceting.

Badoureau was mentioned above. Other investigators include Hess (1878), who discovered two new uniform polyhedra. (Notice the earlier date.) Pitsch (1881) working independently discovered eighteen, some of them not contained in the list by Badoureau. Then between 1930 and 1932 Coxeter and Miller discovered twelve other uniform polyhedra not previously known, but publication was put off in the hope of obtaining a mathematical proof that there are no more. Independently M. S. Longuet-Higgins and H. C. Longuet-Higgins between 1942 and 1944 rediscovered eleven of these twelve. These two teams learned of each others work in 1952. Meanwhile Lesavre and Mercier (1947) rediscovered five of the twelve. In *Uniform polyhedra*, published 1954, from which these facts have been culled, the total now stands at seventy-five uniform polyhedra. But here it is admitted: 'it is the authors' belief that the enumeration is complete, although a rigorous proof has still to be given' (p. 402).

The method used by these recent investigators differs from that of the previous ones. It is based on a systematic investigation of all possible Schwarz triangles as they apply to the polyhedral kaleidoscope. Schwarz triangles are related to the Möbius triangles mentioned previously (see pp. 4–6).

Examples of how metrical properties are found in non-convex uniform polyhedra

Polygons found as faces: {3}, {4}, {5}, {6}, {8}, {10}, to which you must now add the three star polygons: $\{\frac{5}{2}\}$, $\{\frac{8}{3}\}$, $\{\frac{10}{3}\}$

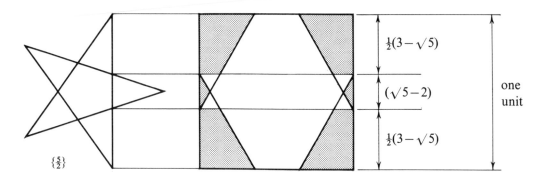

$\{\frac{5}{2}\}$

$\frac{1}{2}(3-\sqrt{5})$

$(\sqrt{5}-2)$

$\frac{1}{2}(3-\sqrt{5})$

one unit

cf. 76

Fig. 34

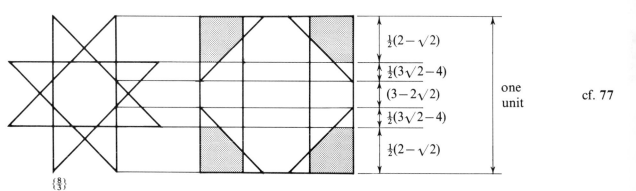

$\{\frac{8}{3}\}$

$\frac{1}{2}(2-\sqrt{2})$

$\frac{1}{2}(3\sqrt{2}-4)$

$(3-2\sqrt{2})$

$\frac{1}{2}(3\sqrt{2}-4)$

$\frac{1}{2}(2-\sqrt{2})$

one unit

cf. 77

Fig. 35

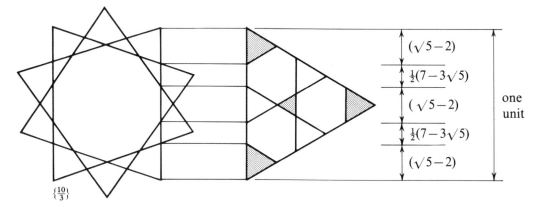

$\{\frac{10}{3}\}$

$(\sqrt{5}-2)$

$\frac{1}{2}(7-3\sqrt{5})$

$(\sqrt{5}-2)$

$\frac{1}{2}(7-3\sqrt{5})$

$(\sqrt{5}-2)$

one unit

cf. 81

Fig. 36

General instructions for making models of non-convex uniform polyhedra

The non-convex uniform polyhedra are described, each showing the facial planes required and the pattern of parts arising from the intersection of the facial planes. No specific dimensions or measurements are given for any of these drawings because there is a very simple relationship which will give you the key to all of them. It lies in the fact that the pentagram or five-pointed star and the decagram or ten-pointed star both exhibit the golden ratio in their dimensions. Thus whenever these stars are found along with other regular polygons these polygons have their edges divided according to the golden ratio $\tau = 1\cdot618$ approximately. This is illustrated by way of the examples on p.99. The octagram or eight-pointed star exhibits the famous $\sqrt{2}$ in its metrical properties, $\sqrt{2} = 1\cdot414$ approximately. Thus once you have drawn these three stars accurately you have all the measurements in the line segments of the stars themselves. It is only in the more intricate models that some further points on the edges will be needed, but again you will find that the golden ratio turns up once more in these smaller segments. Thus a careful study of the drawings will enable you to make the models of any desired size.

The facial planes of the non-convex uniform polyhedra are not always entirely visible. Sometimes some portion of the plane is hidden in the interior of the solid, or a portion is visible as the upper part of the surface while another is visible as the lower or reverse side of the same surface. The light shading is used to show the upper surface, the dark the lower surface, while the invisible portion of the facial plane is left unshaded. When the entire polygon is exterior it is left unshaded. The nets for constructing the models are derived from these visible and invisible portions.

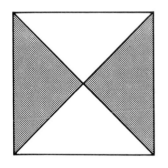

67 Tetrahemihexahedron

This simple polyhedron is easily recognizable as a faceted form of the octahedron. Topologically it is the famous one-sided heptahedron, homomorphic with the one-sided surface named after Steiner. (See Cundy and Rollett, p. 193.) In this polyhedron three equatorial squares lie in three perpendicular planes sharing their edges with four triangles.

To construct a model of this polyhedron four colours may be used. The equilateral triangles may all be the same colour, say R. Cement the isosceles right triangles to the edges of the equilateral triangle as shown, and make four of these units, all with the same colour arrangement. These parts must then be given the form of triangular pyramids, the R triangle serving as the base and the Y, B, O triangles as slant sides. Now a special cementing technique must be employed. Some of the tabs must be turned outward and cemented to form a tongue running along the slant edges of the pyramid, while other tabs are turned inward as usual but left uncemented to form a groove into which the tongue tab of another part can be inserted. If you remember while assembling the parts that each square in the completed model must be the same colour, you can join two appropriate pyramids by applying cement on both sides of a tongue tab before inserting it into the groove of the other part. When you have done this you should have two half-squares, whose planes

$$\tfrac{3}{2}\,3\,|\,2 = r'\,\{\tfrac{3}{3}\}$$
$$4\{3\} + 3\{4\}$$
$$\sqrt{2}$$

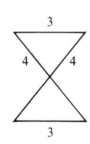

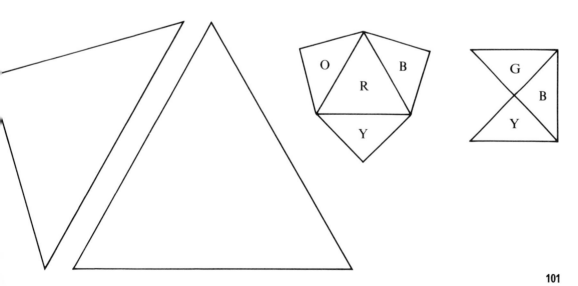

bisect each other at right angles along the line of the tongue and groove edge. The third pyramid may now similarly be cemented in place, and finally the fourth pyramid. You must exercise your own judgement on which tabs to use as tongues and which as grooves. Once you see the model taking shape this will not be hard to do.

An alternative method of construction, also useful for other models, is to make four trihedral cups as shown. All the tabs are turned outward to form ribs on the outside of the cup, which is actually a triangular pyramid without its base. These ribs can then be properly trimmed and manœuvred to serve as double thickness tabs to join the cups together so that an edge of one cup may be made to coincide with that of another. The four R triangles are then added last, cementing one edge at a time and then closing it like a lid in the usual manner. You will find that the acute dihedral angles at the edges make it an easy matter to cement these last equilateral triangles, although your work on the cementing of the cups must be very accurate to make these last triangles fit well. From this point of view the first method of construction is probably better or easier to execute. Your own experience will tell you which one you may prefer.

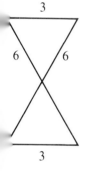

68 Octahemioctahedron

This polyhedron is a faceted cuboctahedron, also called an octatetrahedron. Four equatorial hexagons share their edges with eight triangles. Here again two methods of construction are possible. You can make eight tetrahedra, the first four having the colour arrangement set out below.

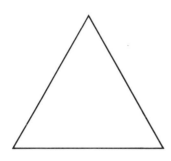

```
1  2  3  4
B  Y  O  G
B  O  Y  R
B  G  O  R
B  Y  G  R
```

$$\tfrac{3}{2}\ 3\,|\,3$$
$$8\{3\} + 4\{6\}$$
$$2$$

The other four also will all have a B triangle as base, but they will have the enantiomorphous arrangement of side faces. Actually this amounts to reversing only two colours, namely 2 and 3 become 3 and 2. These tetrahedra are then joined to one another by the tongue and groove technique. Again you must exercise your own judgement to decide which tabs to turn in and which to turn out. The completed model will have all outer triangles B and the hexagon planes Y, O, R, G. If you keep this in mind while you are cementing the parts you should have no difficulty in arranging them in the proper positions.

The alternative method of construction is to make six tetrahedral cups, the first three in the colour arrangement shown.

```
1  2  3  4
Y  G  R  O
Y  R  O  G
Y  R  G  O
```

The other three are again enantiomorphous to the first set of three. As before in this method of construction all the tabs will be exterior to the cups. Thus they can serve as double thickness tabs, and be suitably trimmed and disposed so that they can be cemented to join the cups along common edges. Then all the B triangles are added last of all.

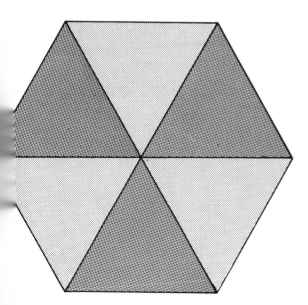

69 Small cubicuboctahedron

This polyhedron is a faceted version of the rhombicuboctahedron. The squares lie in the facial planes of a cube, the octagons lie on parallel planes below the squares, and the triangles are the same as those of the rhombicuboctahedron. Since the cube needs only three colours, a very effective colour arrangement can be obtained here by using the other two colours for the triangles. Use the square, rectangle and two triangles shown in the facial planes for nets. Begin the construction of this model by making four triangular pyramids as set out opposite.

$$\tfrac{3}{2}\ 4\,|\,4$$
$$8\{3\} + 6\{4\} + 6\{8\}$$
$$\sqrt{(5 + 2\sqrt{2})}$$

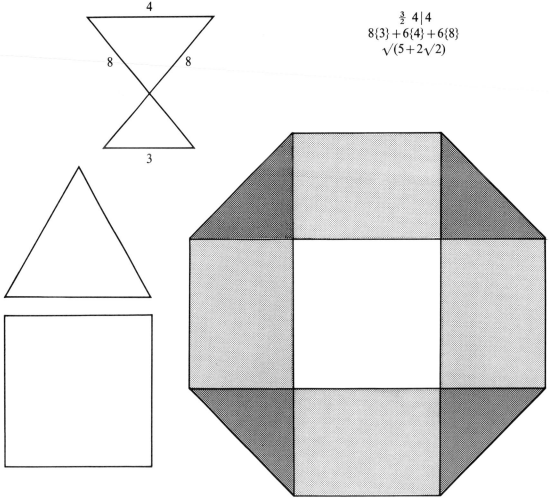

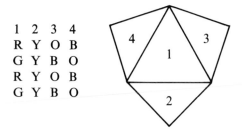

```
1  2  3  4
R  Y  O  B
G  Y  B  O
R  Y  O  B
G  Y  B  O
```

Cement all the tabs outward on the slant edges to form tongues. Next construct an open prism as shown below. This is the upper part of the model, so the B and O rectangles must be turned down in the first part, and the other rectangles treated in a similar fashion. The four pyramids can now be cemented at the four corners of the prism, the tabs between the B and O rectangles forming the groove to receive the tongue tabs of the pyramids. You must of course orient them according to their appropriate colours, keeping the Y triangles below. Now add a ring of four more prisms. They are arranged as set out in the second and third line of the colour table that follows, two of each.

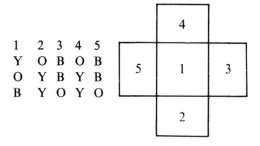

```
1  2  3  4  5
Y  O  B  O  B
O  Y  B  Y  B
B  Y  O  Y  O
```

The remaining four pyramids and three prisms are identical in colour arrangement and the technique of assembly is the same. Actually you will discover that the rectangles from the prisms can simply be turned into place so that the tabs on their edges easily make contact with the tongue tabs of the pyramids without worrying about the groove arrangement. The end result is equivalent to the groove but it is achieved in a different way. This will become clear as the model takes shape. This turns out to be a very sturdy and attractive model.

70 Small ditrigonal icosidodecahedron

This polyhedron has twelve pentagrams in the same facial planes as the dodecahedron and twenty triangles in those of the icosahedron. As is evident from the vertex figure they meet in alternate sets of three around each vertex of the polyhedron. So it can be called a ditrigonal icosidodecahedron. The pentagrams may be done in six colour pairs, but to preserve the map colouring principle you will have to use the second or alternative icosahedral arrangement for the triangular planes. This can be achieved most easily by working around each star, cementing the smaller triangular pairs as dihedral grooves between the star arms. The arrangement and colour table are set out opposite. As soon as you have completed the grooves around the W star, you can immediately add the next five coloured stars. To get these correctly placed, put the G star opposite the Y triangle, and so on around with Y, B, O, R, G.

$$3 \mid 3 \tfrac{5}{2}$$
$$20\{3\} + 12\{\tfrac{5}{2}\}$$
$$\sqrt{3}$$

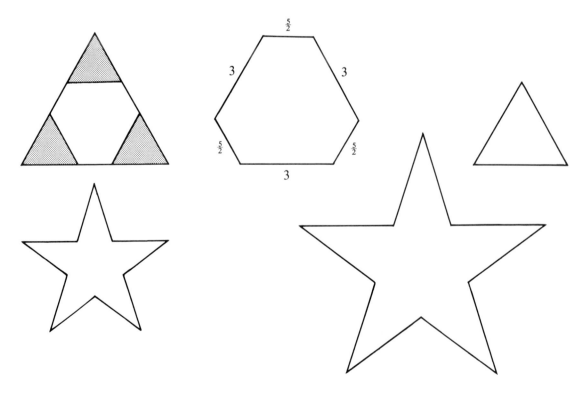

Star	01	23	45	67	89
W	GY	RG	OR	BO	YB
Y	GB	OG	GO	RG	BR
B	YO	RY	YR	GY	OG
O	BR	GB	BG	YB	RY
R	OG	YO	OY	BO	GB
G	RY	BR	RB	OR	YO

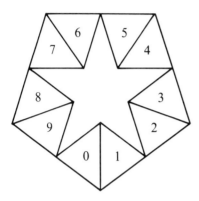

In the colour table it is evident that each colour pair is named twice. This makes it simpler to follow, because you can actually work in that order, surrounding each star with the dihedral grooves. Undoubtedly you have by now observed the cyclic permutation of colours. This is also evident in the colour table if, disregarding the first line, you read down each column successively.

Once this much of the model has been completed you should have no more difficulty. The remaining pentagrams should be placed in opposite colour pairs and the grooves are all determined by watching the triangular planes to see that they are kept in their colour sequence. You will find it easier to cement these triangle pairs together first, and then to cement them between the star arms. All the remaining triangles can be completed in this way. Thus a W star is cemented last of all.

This is best done in stages; cement only one tab first and let this set up firmly, then work on one edge at a time. As the openings at the edges begin to narrow, apply the cement carefully and work it in with the probing needle. Deft fingers and a little patience will do the rest.

71 Small icosicosidodecahedron

In this polyhedron the twenty triangles are on facial planes above and parallel to twenty hexagons. This means that the first icosahedral colour arrangement will do very well here with its five colours, parallel planes being the same colour, leaving white for all the stars. The triangles have edges in common with the hexagons, and vertices in common with the stars. This leaves grooves between the star arms, each groove being formed of two trapezia coming from intersecting hexagon planes which suggests the following method of assembly. Make five grooves as illustrated below following the (0) line in the colour table. These are cemented so that they radiate outward and downward between the W star arms. Triangles are then cemented between the grooves, each colour being determined by the hexagon plane below it. In the colour table each groove is mentioned twice, but again this is easier

$$3 \tfrac{5}{2} \mid 3$$
$$20\{3\} + 20\{6\} + 12\{\tfrac{5}{2}\}$$
$$\sqrt{\dfrac{17 + \sqrt{35}}{2}}$$

	12	12	12	12	12
(0)	YG	BY	OB	RO	GR
(1)	YG	RY	BR	OB	GO
(2)	BY	GB	OG	RO	YR
(3)	OB	YO	RY	GR	BG
(4)	RO	BR	GB	BG	OY
(5)	GR	OG	YO	BY	RB

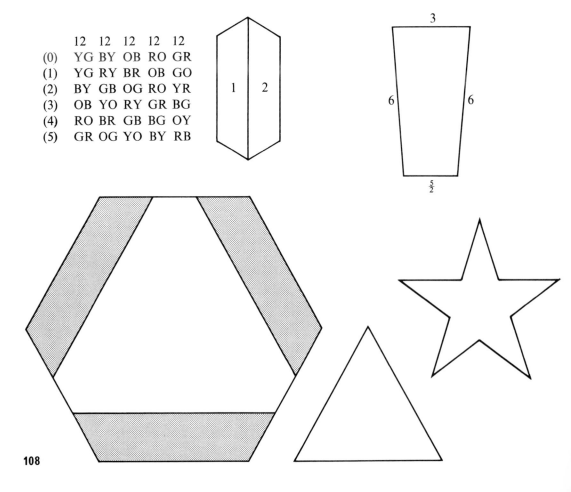

to follow from the point of view of construction. If you work systematically around each star the colour arrangements are easy to follow and you can cement the grooves in place as you determine their colour and position. The grooves and triangles thus help to determine each other as far as colour is concerned. You will find that opposite grooves on the model are enantiomorphs, but in cementing the trapezia pairs it does not matter too much. Turning the pairs end for end will reverse the colours as required.

This model requires a bit of patience to complete, since each star with its ten edges has ten trapezia joined to it, that is, five grooves. The secret is not to try to do too much at once. One edge cemented at a time is the best rule to follow. If you keep five clamps handy, you can keep moving them around and by the time you finish cementing the tabs at a fifth edge you can remove the clamp from the first edge. Also it is generally easier to cement the longer edges of the trapezia first to the triangles which they surround and then the shorter edges or ends of the grooves to the star arms.

72 Small dodecicosi-dodecahedron

This polyhedron is easily recognizable as a faceted rhombicosidodecahedron. The six colour dodecahedral arrangement can be used for the pentagons and the decagons which lie on parallel planes one above the other. Thus the procedure for constructing this model is to begin with a W pentagon to which five trapezia are cemented as shown. Turn the tabs on these trapezia inward but leave them uncemented to serve as grooves into which the tongue tabs of the small triangular pyramids must be inserted. The colour tables for both these parts are set out below.

You must now follow the second icosahedral arrangement to place the triangular pyramids in their correct positions. The colour tables below give only half the required parts, but again the rest are all enantiomorphous to these and they each take their positions diametrically opposite their counterparts. In this way the map colouring principle is preserved.

$$\frac{3}{2}\ 5\,|\,5$$
$$20\{3\} + 12\{5\} + 12\{10\}$$
$$\sqrt{(11 + 4\sqrt{5})}$$

	0	1	2	3	4	5
(0)	W	Y	B	O	R	G
(1)	Y	W	G	O	R	B
(2)	B	W	Y	R	G	O
(3)	O	W	B	G	Y	R
(4)	R	W	O	Y	B	G
(5)	G	W	R	B	O	Y

	0	1	2	3
(1)	Y	W	O	R
(2)	B	W	R	G
(3)	O	W	G	Y
(4)	R	W	Y	B
(5)	G	W	B	O
(6)	Y	G	O	B
(7)	B	Y	R	O
(8)	O	B	G	R
(9)	R	O	Y	G
(10)	G	R	B	Y

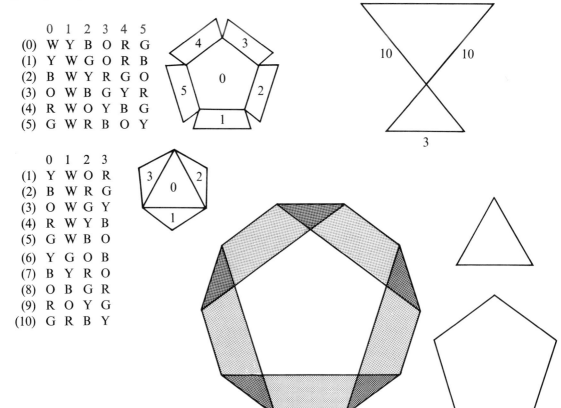

110

73 Dodecadodecahedron

This polyhedron has twelve stars in the same facial planes as the dodecahedron, but on parallel planes below these stars are twelve pentagons each sharing edges with five surrounding stars and intersecting each other. To construct a model of this polyhedron, begin by cementing a set of five trihedral dimples between the arms of a W star. The arrangement is shown in the colour table.

The first set of five dimples are cemented so that the rhombus 1 is below the W star, forming a white pentagon plane parallel with the star. The other colours begin the next five intersecting pentagon planes so that two rhombi appear with the same colour along the straight line between alternate star arms. The next five stars may now be cemented in place, their colours being determined by the two rhombi just mentioned. The next set of five dimples are then added, (6) below (1), and so on for the rest. Enantiomorphism applies for the remaining parts.

$$2\,|\,\tfrac{5}{2}\ 5 = \begin{Bmatrix} \tfrac{5}{2} \\ 5 \end{Bmatrix}$$
$$\frac{12\{\tfrac{5}{2}\} + 12\{5\}}{2}$$

	1	2	3
(1)	W	Y	G
(2)	W	B	Y
(3)	W	O	B
(4)	W	R	O
(5)	W	G	R
(6)	O	G	Y
(7)	R	Y	B
(8)	G	B	O
(9)	Y	O	R
(10)	B	R	G

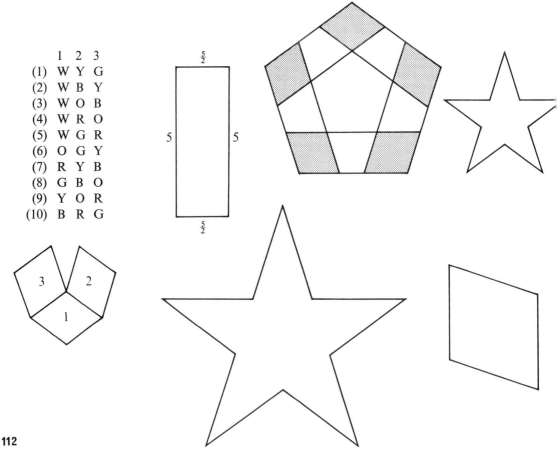

74 Small rhombidodecahedron

This polyhedron is another version of the rhombicosidodecahedron. Here the pentagons are removed giving place to shallow pentagonal cups whose bottoms, also pentagons, belong to the decagon planes. The triangles, as well, are removed leaving shallow dimples whose faces are also part of the decagon planes. The squares, however, are retained. The dodecahedral colour arrangement works well for the decagon planes and suggests the following method of assembly. Begin with a W pentagon and cement five trapezia to it forming a shallow pentagonal cup as shown. You can follow the same colour table as that used for the small stellated dodecahedron. All twelve cups can be cemented together to form an interior dodecahedron. This leaves the spaces between the trapezia to be filled alternately with squares and shallow trihedral dimples. The colour tables set out below give a remarkable result. Cement the Y square between the W and R pentagons, with the other squares following round in order, the B square between the W and G pentagon, and so on. This gives a ring of squares at the top in the usual order: Y, B, O, R, G. The next five rings are: (1), (2), (3), (4), (5).

$$2\ 5\ \begin{vmatrix} \frac{3}{2} \\ \frac{5}{2} \end{vmatrix}$$

$$30\{4\} + 12\{10\}$$

$$\sqrt{(11 + 4\sqrt{5})}$$

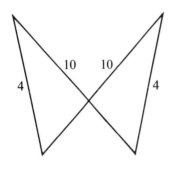

Pentagons	Squares				
(0) W	Y	B	O	R	G
(1) R	Y	B	G	W	O
(2) G	B	O	Y	W	R
(3) Y	O	R	B	W	G
(4) B	R	G	O	W	Y
(5) O	G	Y	R	W	B

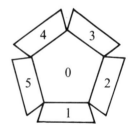

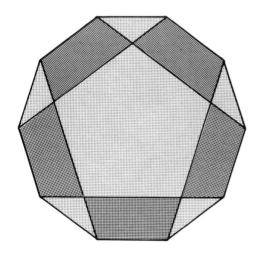

You will note that the rings have squares in common, so each square is listed twice in the table. Again enantiomorphism applies to the remaining rings of squares. The remarkable result mentioned above can now be observed. The five W squares alternate with the other five coloured squares as an equatorial skew band when the polyhedron is held so that the two W decagons are at the poles. The same relationship holds for each of the six colours.

The trihedral dimples have the same colour arrangement and even the same shape as those of the great dodecahedron. However, because of the shallowness it is advisable to eliminate the tabs altogether by cementing the tab of one triangle directly to the under surface of another as shown. Their positions are not hard to locate on the model.

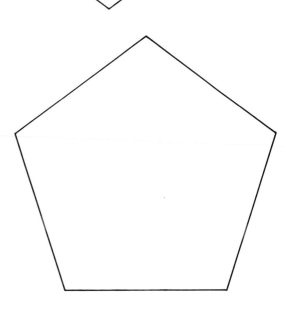

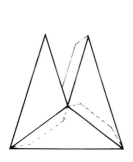

75 Truncated great dodecahedron

The same colour arrangement may be used here as for **21**. The stars and decagons being on parallel planes should be the same colour. The best method for constructing this model is simply to make the trihedral dimples following the colour table for the great dodecahedron. Cement these dimples together along their remaining long edges and add stars at the short edges as required. The arrangement for one dimple is shown.

$$2 \tfrac{5}{2} \mid 5 = t\{5, \tfrac{5}{2}\}$$
$$12\{\tfrac{5}{2}\} + 12\{10\}$$
$$\sqrt{\frac{17 + 5\sqrt{5}}{2}}$$

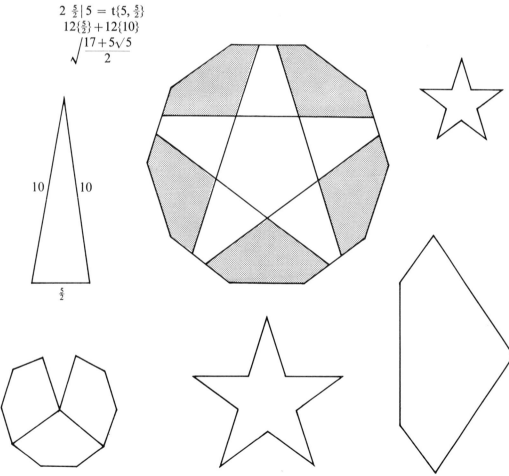

76 Rhombidodecadodecahedron

This lovely polyhedron is almost as spherical as a beach ball and with the arrangement suggested here is equally as colourful. Its name suggests its relation to earlier models. This model has a great number of parts, a total of 312, to cut, trim, and cement. The usual technique of providing all the parts with tabs all around will produce a fairly good model, if it is not too large. The squares can be arranged as in **74**, but here the planes intersect each other so that the skew band is all the more delightful. An arrangement of parts and a colour table are set out opposite to help you get started.

With these parts cemented you will have no trouble finding the right colours for the smallest parts belonging to the pentagon planes. These are cemented in place at once. Continue to work on the rest of the model and complete it except for the small triangular holes that will be left. These holes are closed with small shallow trihedral dimples whose three triangles come from the square planes. Again the colours for these are now not hard to determine. One extra word of advice. It is usually easier to cement concave parts together first, then to cement them to the model as you near the end of your work. The triangles 11–15 thus become folded rhombi in two colours and serve as connectors between sections.

$$\tfrac{5}{2}\ 5\,|\,2 = r\begin{Bmatrix}\tfrac{5}{2}\\5\end{Bmatrix}$$

$$12\{\tfrac{5}{2}\} + 30\{4\} + 12\{5\}$$

$$\sqrt{7}$$

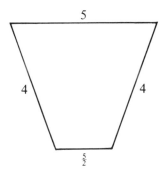

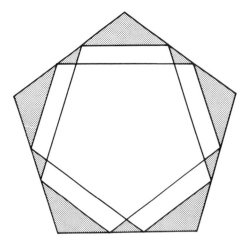

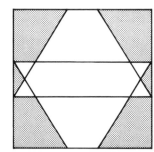

0	1	2	3	4	5	6	7	8	9	10	11	12	13	14	15
W	Y	G	B	Y	O	B	R	O	G	R	O	R	G	Y	B
Y	W	B	G	W	O	G	R	O	B	R	O	R	B	W	G
B	W	O	Y	W	R	Y	G	R	O	G	R	G	O	W	Y
O	W	R	B	W	G	B	Y	G	R	Y	G	Y	R	W	B
R	W	G	O	W	Y	O	B	Y	G	B	Y	B	G	W	O
G	W	Y	R	W	B	R	O	B	Y	O	B	O	Y	W	R

117

77 Great cubicuboctahedron

This polyhedron is a faceted cube. Each octagram lies on the face of a cube which you may imagine as enclosing the polyhedron. Then each corner gets a tetrahedral dimple and each edge gets a dihedral groove. The dimples and grooves alternate between the star arms. Since the cube can be done with three colours the octagrams can follow the same arrangement. Six stars are thus paired and six squares get the same colours because they are parallel to the stars and below them. The triangles can then use the other two colours alternately.

Begin constructing this model with a Y octagram. Then make four dimples as shown, also four triangle pairs. Note that these triangles are slightly larger than those used in the dimples. Cement the dimples and grooves alternately between the star arms, seeing that the colours run on their respective planes.

With this done you should have no further difficulty. The colours are easily determined for the stars, the second Y star being cemented last, one edge at a time.

$$3 \ 4\,|\,\tfrac{4}{3}$$
$$8\{3\} + 6\{4\} + 6\{\tfrac{8}{3}\}$$
$$\sqrt{(5 - 2\sqrt{2})}$$

0	1	2	3		1	2
G	Y	B	O		R	G
R	Y	O	B		G	R
G	Y	B	O		R	G
R	Y	O	B		G	R

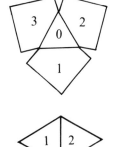

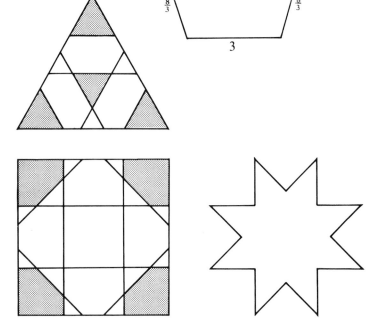

78 Cubohemioctahedron

This polyhedron is another faceted version of the cuboctahedron and thus the same construction methods are applicable here as in **68**. Prepare your nets from the facial planes. You may make six pyramids with square bases, three as set out below and the other three their enantiomorphs. Assembly of these pyramids is by the tongue and groove technique.

The alternative method of assembly has eight trihedral cups, the first four arranged as shown and the other four as their enantiomorphs. As usual all tabs are turned outward, then the double tab of one cup is cemented to the under-surface of another to make the edges coincide. The squares are added last to complete the model.

$$\tfrac{4}{3}\ 4\,|\,3$$
$$4\{6\} + 6\{4\}$$
$$2$$

0	1	2	3	4
B	Y	O	R	G
B	Y	R	O	G
B	Y	R	G	O

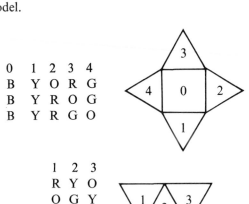

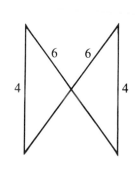

1	2	3
R	Y	O
O	G	Y
R	G	O
Y	G	R

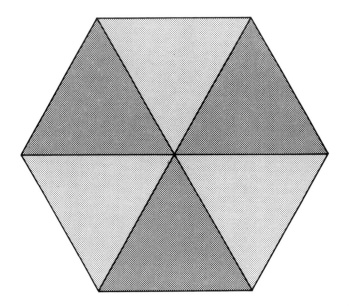

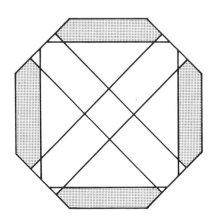

79 Cuboctatruncated cuboctahedron

This polyhedron is a faceted octahedron. The dihedral grooves between the star arms are parts of hexagon planes. The relationship to the octahedron suggests a very effective colour arrangement. The eight hexagons can be parallel pairs in four colours, and all six octagrams are a fifth colour. The stars may all be white or the same colour as the octagrams since they are on planes parallel to and above the octagram planes. To construct a model of this polyhedron, begin by surrounding triangles, the central part of the hexagon, with the parts coming from the octagon planes as shown. The shortest edges of these surrounding parts should then be cemented to form a shallow cup. Next the grooves are made as shown and these two parts cemented alternately around an octagram.

$$3 \ \tfrac{4}{3} \ 4|$$
$$8\{6\} + 6\{8\} + 6\{\tfrac{8}{3}\}$$
$$\sqrt{7}$$

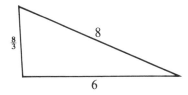

0	1	2	3	4	5
Y	G	G	G	Y	R
B	G	G	G	B	Y
O	G	G	G	O	B
R	G	G	G	R	O

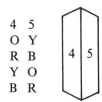

The next four octagrams can be added immediately. After this four more grooves form an equatorial band connecting these octagrams. Their colours are

4	5
O	Y
R	B
Y	O
B	R

You will have noted that the last two of these pairs are enantiomorphs of the first two. They appear of course diametrically opposite on the model. The next set of four shallow cups and their four connecting grooves may then be

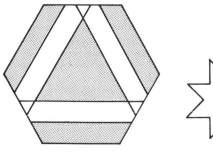

121

cemented in place. The cups are each opposite their respective colours and the grooves opposite their enantiomorphs. In cementing the pairs that form the grooves you need not worry about right or left because turning this part end for end will give you the desired order. A final octagram then completes the model. This makes an attractive and rigid polyhedron.

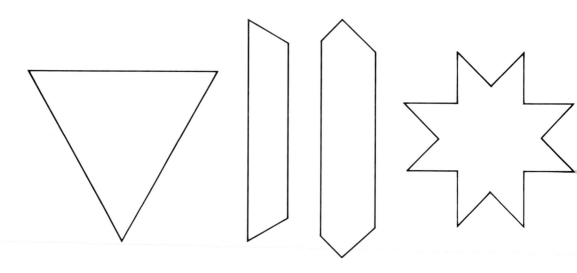

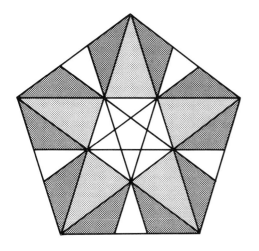

80 Ditrigonal dodecahedron

This polyhedron is especially interesting because of its close relationship to the great stellated dodecahedron. It may be thought of as this latter solid with the embossed stars of the great dodecahedron removed from their pentagonal planes, then turned over and set down to make the vertices of these embossed stars coincide with the points between five vertex parts of the great stellated dodecahedron **22**, the edges being shared in common. This sounds a bit complicated in words, but once the model is completed you will readily see it for yourself. This model can also be thought of as a faceted version of **70**. The dihedral triangular grooves of **70** are here removed to give place to deep holes between the star arms. These holes are formed by the intersecting pentagon planes, and thus their faces are the familiar 72°, 36° and 36° 108° isosceles triangles. The relationship of this polyhedron to the two stellations of the dodecahedron means that the same colour arrangement is suitable here.

To make a model of this polyhedron, prepare twelve pentagrams, two of each of the six colours. Next prepare a set of five hexahedral cups, the holes described above, whose parts are shown on p. 124. A complete colour table for two more sets of five is set out on the same page.

Turn the tabs to the outside of these cups, which are then cemented between the star arms. The first set of five cups will surround the W pentagram. After these are in place add the five coloured pentagrams. The Y star must lie on a plane above the Y pentagon, which shows only two parts of its area in the cups. These are the

$$3\left|\tfrac{5}{3}\right.5$$
$$12\{\tfrac{5}{2}\} + 12\{5\}$$
$$\sqrt{3}$$

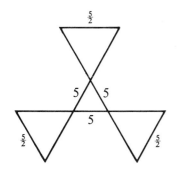

triangles 1 or 2. Once you locate the correct position for this first star the others follow around in order as Y, B, O, R, G. Then the second set of five cups can be prepared and used as connectors. Cup (6) goes between the G and Y star, (7) between the Y and B star, and so on around in order. The third set of five cups will complete half the model. Their positions are easily found by watching the pentagon planes to see that they keep their respective colours. The second half of the model is enantiomorphous to the first, the parts being diametrically opposite their counterparts. The W star is cemented last. This makes a very sturdy model, but as you approach the end of your work it requires very careful cementing to make the stars fit well.

	1	2	3	4	5	6
(1)	B	G		O	R	O	R
(2)	O	Y		R	G	R	G
(3)	R	B		G	Y	G	Y
(4)	G	O		Y	B	Y	B
(5)	Y	R		B	O	B	O
(6)	W	O		R	B	R	B
(7)	W	R		G	O	G	O
(8)	W	G		Y	R	Y	R
(9)	W	Y		B	G	B	G
(10)	W	B		O	Y	O	Y
(11)	G	R		B	W	B	W
(12)	Y	G		O	W	O	W
(13)	B	Y		R	W	R	W
(14)	O	B		G	W	G	W
(15)	R	O		Y	W	Y	W

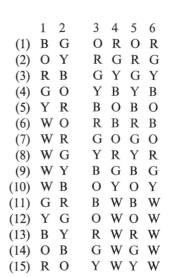

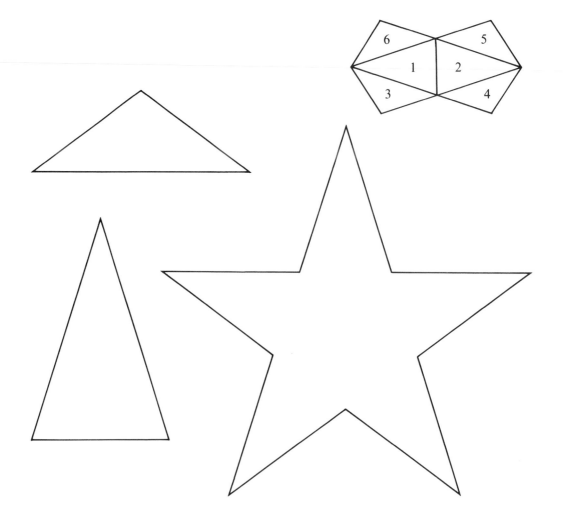

124

81 Great ditrigonal dodec- icosidodecahedron

This polyhedron is a member of the icosidodeca-hedral family. This suggests the usual colour arrangement, six colours in opposite pairs for the decagrams and the pentagon planes parallel to them, and the second five-colour icosahedral arrangement for the triangles, making opposite triangular planes the same colour.

The simplest method of assembly for a model of this polyhedron is to make the dimples and grooves and cement them alternately between the star arms. Start with a W decagram cementing the parts as shown, and continue in the same way to complete the model.

$$3 \ 5|\tfrac{5}{3}$$
$$20\{3\} + 12\{5\} + 12\{\tfrac{10}{3}\}$$
$$\sqrt{\dfrac{17 - 3\sqrt{5}}{2}}$$

1	2	3	4	5	6
Y	W	O	R	B	Y
B	W	R	G	O	B
O	W	G	Y	R	O
R	W	Y	B	G	R
G	W	B	O	Y	G

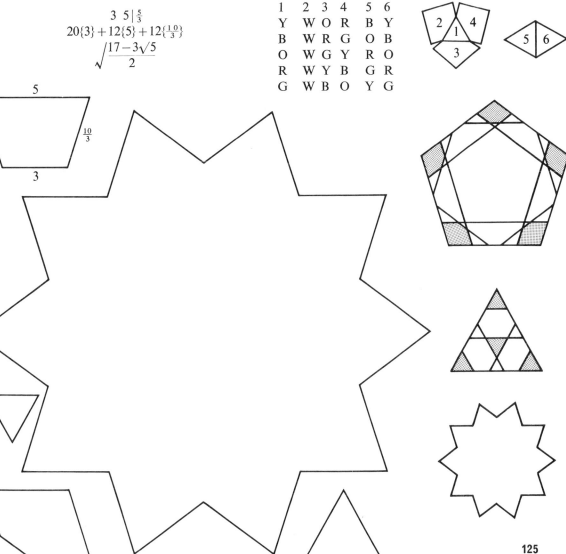

82 Small ditrigonal dodec-icosidodecahedron

This is a most remarkable polyhedron because it was discovered in our own century and published for the first time in 1954 (see Coxeter, *op. cit.*), yet it is closely related to the stellated dodecahedra. From this point of view it is strange that it should have been missed by the earlier investigators. But then the stellated dodecahedra themselves were missed until Kepler's time. This polyhedron can be thought of as a truncated form of the great stellated dodecahedron with the embossed or solid stars of the great dodecahedron taken from their planes and turned over here to fill the spaces between the stumps remaining after truncation. This relationship is helpful in making a model of this polyhedron.

Make twenty truncated pyramids as shown. The side faces follow the colour arrangement of the great stellated dodecahedron. Leave the tabs at the slant edges uncemented but turned in to serve as grooves into which tongue tabs are inserted. The triangles follow the first icosahedral arrangement.

$$3\ \tfrac{5}{3}\,|\,5$$
$$20\{3\} + 12\{\tfrac{5}{2}\} + 12\{10\}$$
$$\sqrt{\frac{17 + 3\sqrt{5}}{2}}$$

	1	2	3	4
	Y	Y	G	B
	B	B	Y	O
	O	O	B	R
	R	R	O	G
	G	G	R	Y

You will at once see that the map colouring principle is abandoned here, but only in the first ring of five parts. Also once you have cemented these parts as a ring, you will see the position at their centre for the first solid star part. These solid stars are made following the colour arrangement of the great dodecahedron. Cement one short edge of the 36°, 108° triangles between the star arms of the pentagram, but turn the tabs on the longer edges out to form tongue tabs. The other short edges are turned in and cemented to each other, thus forming a sort of star pyramid

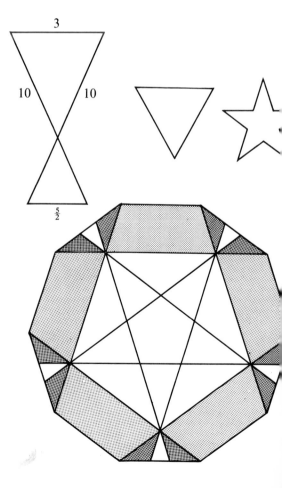

with concave edges as usual, but with ribs on the outside at the convex edges. These solid stars can now be easily cemented into their respective places by inserting the five tongue tabs into the five grooves made by the tabs of the trapezia 2, 3, 4 shown opposite. The W pentagram, call it the base of the solid star pyramid, should lie on a plane above and parallel to the W decagon, whose parts appear alternately with upper and then lower surface visible. The procedure will become apparent as the work proceeds. This makes a very sturdy, rigid, and interesting model.

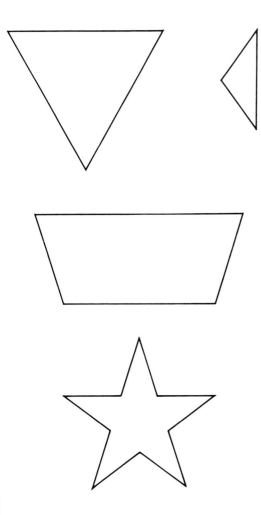

83 Icosidodecadodecahedron

This polyhedron is very much like **76** but the squares there are replaced with hexagons here, giving it a more fascinatingly dimpled character and also adding greater rigidity and beauty. Once you begin to make this model you will be surprised to see how easily it can be assembled despite its complicated structure.

To simplify construction, the procedure has been broken down into four steps or parts. Solid stars, called part I, are built first. Begin with a regular pentagram and fill the spaces between the star arms with triangles belonging to the hexagon plane to make a low inverted star pyramid. The longest tabs on these triangles are turned out to form ribs along the slant edges leading from the star points to the central vertex.

$$\tfrac{5}{3} \; 5\,|\,3$$
$$20\{6\} + 12\{\tfrac{5}{2}\} + 12\{5\}$$
$$\sqrt{7}$$

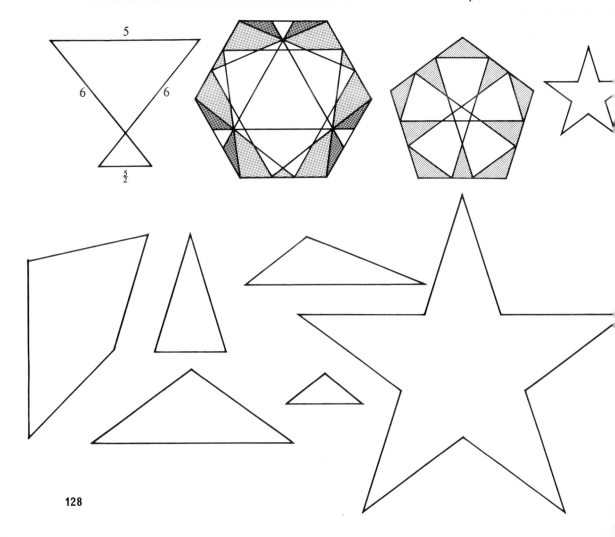

The figure shows a plan of this part and a colour table for six such parts.

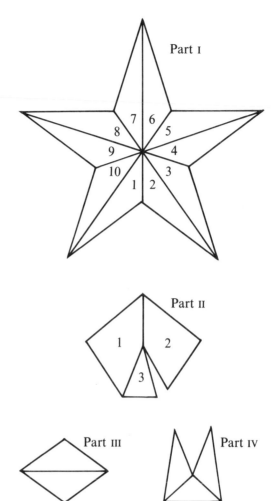

Part I

Star	1	2	3	4	5	6	7	8	9	10
(0)	W	Y	G	B	Y	O	B	R	O	G	R
(1)	Y	B	G	Y	B	R	Y	O	R	G	O
(2)	B	O	Y	B	O	G	B	R	G	Y	R
(3)	O	R	B	O	R	Y	O	G	Y	B	G
(4)	R	G	O	R	G	B	R	Y	B	O	Y
(5)	G	Y	R	G	Y	O	G	B	O	R	B

Part II is a sort of wedge or open trihedral scoop. The sides belong to the hexagon planes and the central triangle comes from the pentagon plane.

The colour table follows:

	1	2	1	2	1	2	1	2	1	2	3
(0)	G	O	Y	R	B	G	O	Y	R	B	W
(1)	Y	G	R	B	O	Y	G	R	B	O	Y
(2)	B	Y	G	O	R	B	Y	G	O	R	B
(3)	O	B	Y	R	G	O	B	Y	R	G	O
(4)	R	O	B	G	Y	R	O	B	G	Y	R
(5)	G	R	O	Y	B	G	R	O	Y	B	G

Part II is cemented so that it shares its acute vertex with the central point of the solid star and its longer edges run along the ribs of the solid star filling the spaces between the star arms. In each case the first-named colour pair of part II goes between sections 7, 8 of part I and the rest follow round in order. If you use clamps, part II can easily be kept in place. In effect you have here the tongue and groove technique but achieved in a different manner. It is this different method of assembly which makes this model surprisingly easy. Parts I and II make a section. Twelve sections complete the model. These sections are joined together using part III as connectors. These are the same colour pairs as in **76**, and their position is now easy to determine. Finally twenty more small trihedral dimples, part IV, are needed to close the small triangular openings at the base of part II. Their colour arrangement also is easily determined from the hexagon planes.

You will notice that the map colouring principle has again been abandoned. You may wish to experiment on your own with more colours to get a better effect, although the arrangement described above is very satisfactory. Only a trained eye would notice the colour defect, if it may be so called. This model takes a great deal of time to assemble because of its many parts. Again you may wish to experiment on your own by making other nets combining more of the parts to save labour. You will have to be the judge of whether the result is more pleasing. Undoubtedly this model is one of the most satisfactory of the whole set of uniform polyhedra from the point of view of being easy to make yet being complex in appearance. It is easy to make from the point of view of assembly, not of time involved. It turns out also to be a very rigid model.

84 Icosidodecatruncated icosidodecahedron

This polyhedron is to the icosahedral group, what **79** is to the octahedral. The construction technique remains the same. To make this model, build shallow triangular cups and dihedral grooves and cement these alternately between the star arms of the decagrams. A suitable colour scheme is given opposite.

The first five of these cups surround the W decagram, the paired dihedral grooves connecting adjacent cups.

The first set of five grooves extends radially downward from between the star arms. The next set of five can now be cemented along their edges, the Y star above the Y decagon plane which at this stage of construction shows only two of its five parts. The other stars follow around in the usual order. Now the third set of shallow cups may be added, (6) below (1), and so on. The colours of the second set of grooves are determined by the triangles at the bottom of the cups. So too with the third set of grooves. Again the map colouring principle has been overlooked, but this must happen when the dodecahedral and icosahedral colour arrangements are both used in one model, while keeping to only six colours. Here the second icosahedral arrangement is used. To complete the model make the remaining parts enantiomorphous to those tabulated. The numerous edges around each decagram require perseverance in cementing the tabs. Also these decagrams may need some backing, to make them stiffer, when the scale of the model is fairly large, in which case the parts are also easier to cement.

$$3 \ \tfrac{5}{3} \ 5 \,|$$
$$20\{6\} + 12\{10\} + 12\{\tfrac{10}{3}\}$$
$$4$$

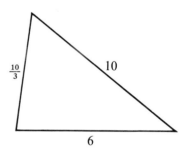

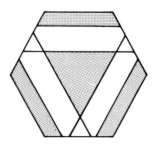

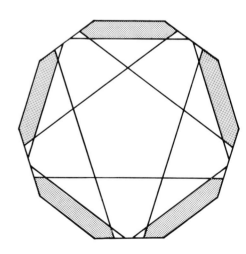

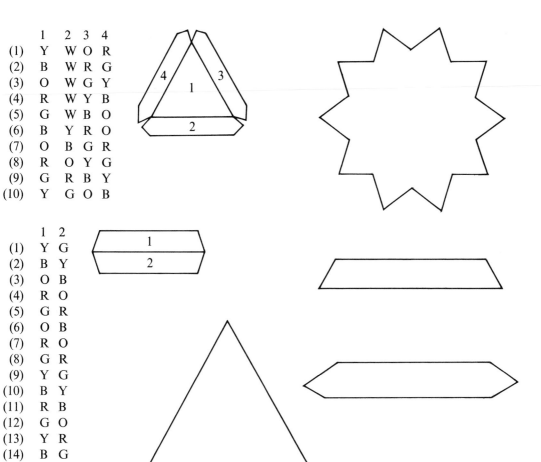

	1	2	3	4
(1)	Y	W	O	R
(2)	B	W	R	G
(3)	O	W	G	Y
(4)	R	W	Y	B
(5)	G	W	B	O
(6)	B	Y	R	O
(7)	O	B	G	R
(8)	R	O	Y	G
(9)	G	R	B	Y
(10)	Y	G	O	B

	1	2
(1)	Y	G
(2)	B	Y
(3)	O	B
(4)	R	O
(5)	G	R
(6)	O	B
(7)	R	O
(8)	G	R
(9)	Y	G
(10)	B	Y
(11)	R	B
(12)	G	O
(13)	Y	R
(14)	B	G
(15)	O	Y

85 Quasirhombicuboctahedron

This polyhedron is very similar to **77**. The octagrams of that polyhedron are removed here with only the edges retained making it deeply dimpled and cupped. It has two different sets of intersecting squares. If you follow the colour arrangement of **77** for the triangles and corner dimples used there, you will again get an effective arrangement. However, with more squares in this model, you cannot preserve the map colouring principle with only five or six colours. In fact the many parts required here call for a great deal of perseverance to complete a model in colour. You may therefore find it more satisfactory to make a model in one colour first, to see what colour possibilities you can work out for yourself.

Suitable nets, combining many parts in one, for such a model are given opposite. The letters identify the individual nets shown separately at full working scale. You will need six of part I, the central portion of the faceted octagram. Leave tabs all round. The broken lines should be scored on the reverse side and the folding along these lines is up instead of down. Cut twenty-four star arms, part II. You will also need eight of part III, the dimpled corner portions, and twelve of part IV. Assemble parts I and II first and then use parts III and IV as connectors.

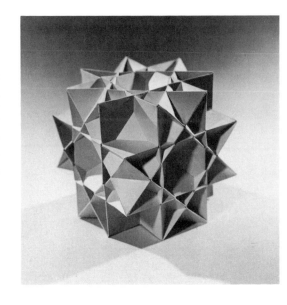

$$\tfrac{3}{2}\,4\,|\,2 = r'\{\tfrac{3}{4}\}$$
$$8\{3\}+(12+6)\{4\}$$
$$\sqrt{(5-2\sqrt{2})}$$

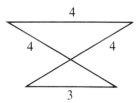

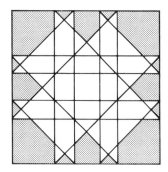

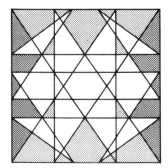

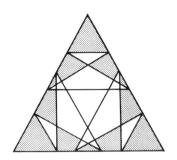

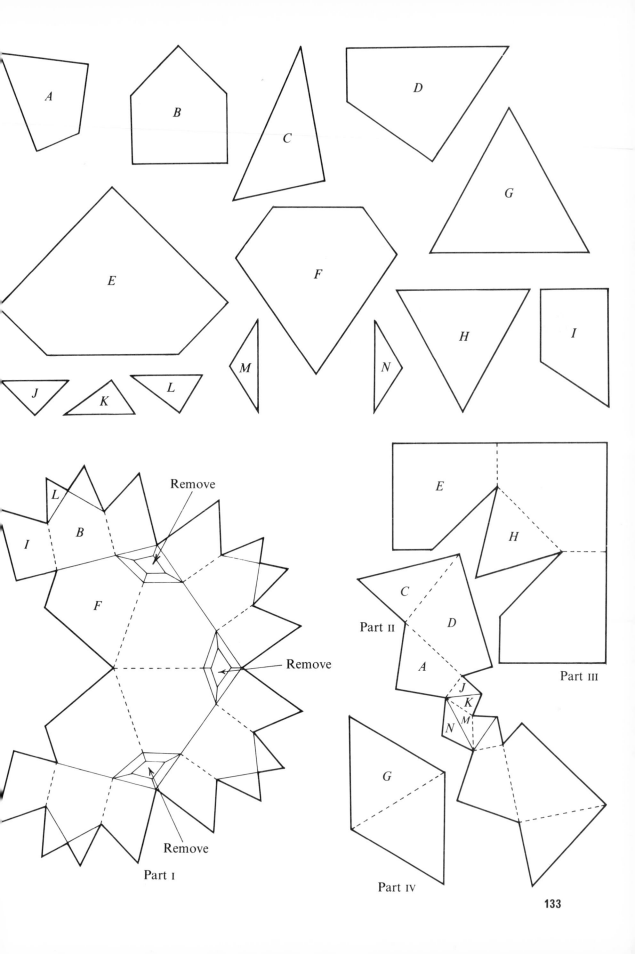

A

B

C

D

G

E

F

H

I

M

N

J

K

L

Remove

L

I

B

F

Remove

Remove

Part I

E

H

C

Part II

D

A

J

K

M

N

G

Part IV

Part III

133

86 Small rhombihexahedron

This polyhedron is another faceted rhombi-cuboctahedron very much like **69**. The triangles found there are here removed and the set of squares found there, are here replaced with a different set of squares. You can use the same set of open prisms in constructing this model as you used in **69**. The triangular pyramids there will now give their sides to trihedral dimples with the same colour arrangement. The square prisms are cemented first to form an interior cube; then the set of squares, which may all be one colour either R or G, are cemented along opposite sides to the edges of the prisms. Finally the dimples are cemented in place. You must of course see to it that the double tabs below the dimples are properly trimmed and adjusted so they do not jam. This makes a very rigid model.

$$2 \ 4 \ \begin{array}{|c} \frac{3}{2} \\ \frac{4}{2} \end{array}$$

$$12\{4\} + 6\{8\}$$

$$\sqrt{(5+2\sqrt{2})}$$

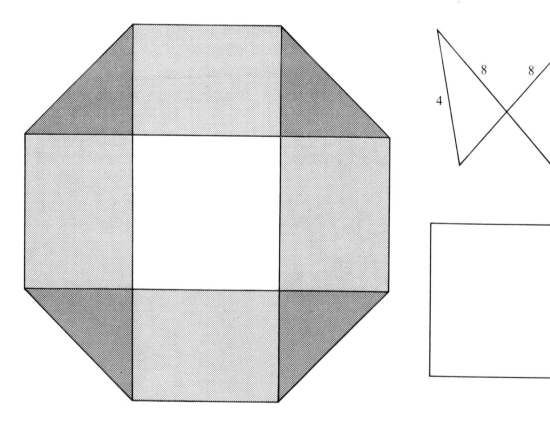

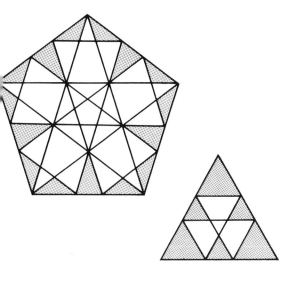

87 Great ditrigonal icosidodecahedron

This polyhedron is another ditrigonal icosidodecahedron like **70**. It differs from **70** in that here the pentagrams are removed and pentagons replace them on a parallel plane closer to the centre of the solid. The twenty triangular planes have exterior parts made up of two sizes of equilateral triangles, and the twelve pentagon planes have the usual 36°, 72° and 36°, 108° isosceles triangles. The colour arrangement can also follow that used in **70**.

To make a model of this polyhedron begin by making a pentahedral cup or inverted pyramid, the same as that used for the small stellated dodecahedron, but turned inwards so the tabs form ribs on the outside of the cup. The same colour scheme works here but turning the parts inside out, you may note, is equivalent to building the enantiomorphous arrangement. Next it will be best to prepare a set of five trihedral dimples, following the colour table on p. 136. These dimples are cemented to the edges or lip of the cup. The structure is a bit unstable at this stage, but if you press this whole section as completed so far against a pane of glass and look through from the other side you will see a perfectly faceted five pointed star. Pairs of equilateral triangles must next be cemented between the star arms, just as in **70**. These begin to give the model some rigidity, but it will be advisable to add more backing inside the star along the edges from point to point. If you make the model without this, you may find these edges are slightly out of line.

The colour arrangement for six faceted stars is shown in the figure and table on p. 136.

The procedure is to continue making the faceted stars and to use the triangle pairs as connectors, as in **70**. Again, the second half of the model is made enantiomorphously. The fact that the map colouring principle is violated is not too noticeable in the completed model. Also it may be easier to follow the model itself as it grows rather than the colour table.

$$\tfrac{3}{2}\,|\,3\ 5$$
$$20\{3\} + 12\{5\}$$
$$\sqrt{3}$$

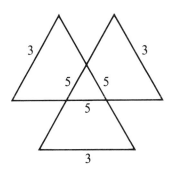

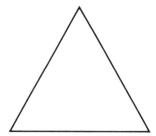

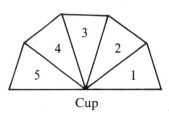

Cup

Cups

	1	2	3	4	5
(0)	Y	B	O	R	G
(1)	W	G	O	R	B
(2)	W	Y	R	G	O
(3)	W	B	G	Y	R
(4)	W	O	Y	B	G
(5)	W	R	B	O	Y

Star arm dimple

Triangle pair

Star arm dimples

	6	7	8		6	7	8
(0)	B	Y	G	(3)	B	Y	R
	O	B	Y		G	R	W
	R	O	B		Y	Y	B
	G	R	O		R	G	G
	Y	G	R		W	O	Y
(1)	G	R	B	(4)	O	B	G
	O	B	W		Y	G	W
	R	R	G		B	B	O
	B	O	O		G	Y	Y
	W	Y	R		W	R	B
(2)	Y	G	O	(5)	R	O	Y
	R	O	W		B	Y	W
	G	G	Y		O	O	R
	O	R	R		Y	B	B
	W	B	G		W	G	O

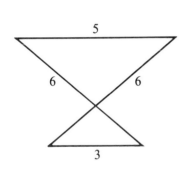

88 Great icosicosidodecahedron

This polyhedron is related to **81**, the difference being that here the decagrams are gone and the edges alone are retained, while the addition of hexagons introduces multifaceted decagrams in place of plane decagrams. And multifaceted is literally true! To make a model of this polyhedron you will have to prepare 76 parts for each faceted decagram alone, not to mention the other parts which serve as connectors. It may interest you to know that the total number of individual small segments of surface area generated by all the intersections of the three regular polygons belonging to the facial planes of this polyhedron reaches the imposing figure of 1232. This is a real challenge to the perseverance of any model maker! Because some of the parts are so small, the model must be on a scale sufficiently large to enable you to handle them. Also

$$\tfrac{3}{2}\ 5\,|\,3$$
$$20\{3\}+20\{6\}+12\{5\}$$
$$\sqrt{\dfrac{17-3\sqrt{5}}{2}}$$

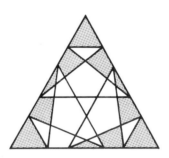

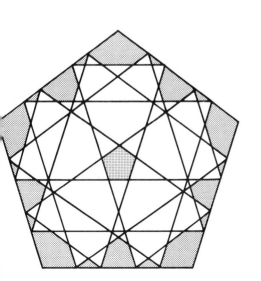

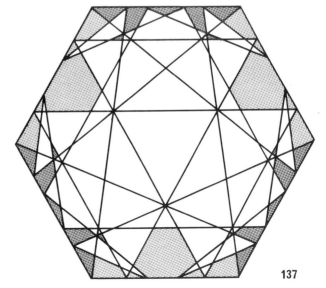

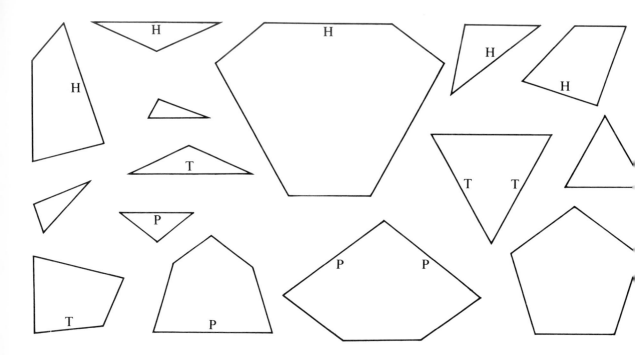

some stiffening is required along the internal decagram edges to keep them straight. The various parts are set out in the figures and a complete colour table is given. You will find it best to keep all pentagon planes W and to use the five colours Y, B, O, R, G for the hexagons and triangles.

Begin with the cup shown in part I. This shows the small triangles needed between 1 and 2, 2 and 3, etc. Part II is a deep cup with a pair of very small triangles at the bottom, which is cemented so the edge of the 0 portion is placed with the edge of 1 2 3 4 5 in part I. The star-arm dimples are shown as part III in enantiomorphic forms since their colour arrangement is not enantiomorphous when they are adjacent. The edges of

2 and 6 of part III are cemented to 1 and 2 of part II. A set of small trihedral dimples will now close the spaces between the star arm pairs and the small triangle mentioned in part I. This last set of dimples is part IV. This completes one faceted decagram. Twelve of these are needed and are joined together with the same dimples and grooves as in **81**, except that here the pentagon planes are all W.

The rest of the model again follows enantiomorphically. All you need is about 30 hours. Once you get working systematically each faceted decagram may take about 2 hours, a total of 24 hours for all, and then another 6 hours to get all parts joined with the paired triangular grooves and corner dimples.

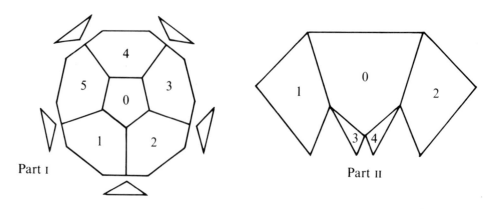

Part I

Part II

Part I						Part II					Part III								Part IV		
0	1	2	3	4	5	0	1	2	3	4	1	2	3	4	5	6	7	8	1	2	3
(0) W	Y	B	O	R	G	W	G	Y	Y	B	W	Y	R	O	W	G	Y	O	O	G	Y
						W	Y	B	B	O	W	B	G	R	W	Y	B	R	R	Y	B
						W	B	O	O	R	W	O	Y	G	W	B	O	G	G	B	O
						W	O	R	R	G	W	R	B	Y	W	O	R	Y	Y	O	R
						W	R	G	G	Y	W	G	O	B	W	R	G	B	B	R	G
(1) W	B	Y	R	O	G	W	G	B	B	Y	W	B	O	R	W	G	B	R	R	G	B
						W	Y	O	O	B	W	O	R	G	W	Y	O	B	G	Y	O
						W	B	R	R	O	W	R	G	Y	W	B	R	Y	Y	B	R
						W	O	G	G	R	W	G	B	Y	W	O	G	B	B	O	G
						W	R	Y	Y	G	W	Y	B	O	W	R	Y	O	O	R	Y
(2) W	O	B	G	R	Y	W	Y	O	O	B	W	O	R	G	W	Y	O	G	G	Y	O
						W	B	R	R	O	W	R	G	Y	W	B	R	Y	Y	B	R
						W	O	G	G	R	W	G	Y	B	W	O	G	B	B	O	G
						W	R	Y	Y	G	W	Y	B	O	W	R	Y	O	O	R	Y
						W	G	B	B	Y	W	B	O	R	W	G	B	R	R	G	B
(3) W	R	O	Y	G	B	W	B	R	R	O	W	R	G	Y	W	B	R	Y	Y	B	R
						W	O	G	G	R	W	G	Y	B	W	O	G	B	B	O	G
						W	R	Y	Y	G	W	Y	B	O	W	R	Y	O	O	R	Y
						W	G	B	B	Y	W	B	O	R	W	G	B	R	R	G	B
						W	Y	O	O	B	W	O	R	G	W	Y	O	G	G	Y	O
(4) W	G	R	B	Y	O	W	O	G	G	R	W	G	Y	B	W	O	G	B	B	O	B
						W	R	Y	Y	G	W	Y	B	O	W	R	Y	O	O	R	Y
						W	G	B	B	Y	W	B	O	R	W	G	B	R	R	G	B
						W	Y	O	O	B	W	O	R	G	W	Y	O	G	B	Y	O
						W	B	R	R	O	W	R	G	Y	W	B	R	Y	Y	B	R
(5) W	Y	G	O	B	R	W	R	Y	Y	G	W	Y	B	O	W	R	Y	O	O	R	Y
						W	G	B	B	Y	W	B	O	R	W	G	B	R	R	G	B
						W	Y	O	O	B	W	O	R	G	W	Y	O	G	B	Y	O
						W	B	R	R	O	W	R	G	Y	W	B	R	Y	Y	B	R
						W	O	G	G	R	W	G	Y	B	W	O	G	B	B	O	G

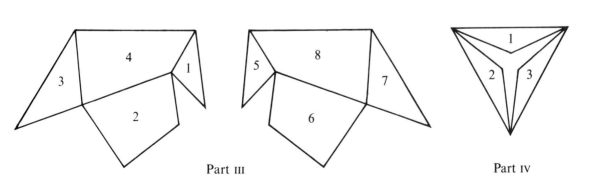

Part III

Part IV

89 Small icosihemidodecahedron

This polyhedron and **91** are both faceted versions of the icosidodecahedron. The decagons cut right through the centre of the solid on equatorial planes. This polyhedron has deep pentahedral cups or inverted pyramids all of which have their apex at the centre of the solid. As for colour arrangements, each of the decagons can have a colour of its own W, Y, B, O, R, G. Then the triangles should follow the usual icosahedral order, namely the second alternative. Again you have a choice of two techniques for construction; the tongue and groove arrangement for twenty triangular pyramids, each of which has the colour arrangement of the great stellated dodecahedron; or the double tabs turned outward as ribs on twelve pentahedral pyramids, as in the small stellated dodecahedron. The triangles are added last of all in this latter method, which is probably the easier to execute. If you try both you can see for yourself which gives the best result. This makes a very rigid model.

$$\tfrac{3}{2}\ 3\,|\,5$$
$$20\{3\} + 6\{10\}$$
$$2\tau$$

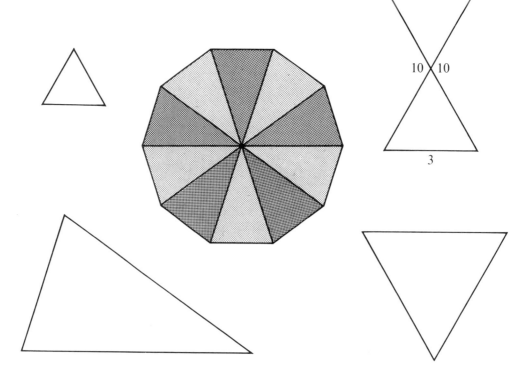

90 Small dodecicosahedron

This polyhedron forms a pair with **82** in that it was also published for the first time in 1954. (See Coxeter, *op. cit.*) Faceted stars are again found here, formed by the intersection of only two types of planes, hexagons and decagons. The first icosahedral arrangement of colours may be used effectively for the twenty hexagon planes and the dodecahedral arrangement for the decagons. The method of assembly for this model is to make the faceted stars whose central facets follow the order of the six colours used in **20**, but forming cups or inverted pyramids. The star arms have dimples made of small equilateral triangles belonging to the hexagons and two isosceles triangles belonging to the decagons. You can get these dimples arranged correctly by following the colours from the cups outward to the star arms, only you must give the equilateral triangles a turn counterclockwise two places to begin with, as shown below. Pairs of trapezia form dihedral grooves between the star arms with the usual arrangement. The shallow triangular cups have their vertices in common with the vertices of the star arms and their side faces continue the facial planes of the side faces of the star arm dimples, so the colours are not hard to match. In the next set of five faceted stars you will find the Y hexagon sharing one of its edges with a Y decagon, and so on around for the other colours. This defect in the map colouring principle again does not seriously affect the end result.

The rest are done in the usual cyclic permutation of colours, with opposite parts enantiomorphous. You should now be able to complete the model without further directions. The faceted stars are not entirely rigid and so some internal stiffening may be required. However, this is not needed if the model is small.

$$3 \ 5 \ \left|\begin{array}{c}\frac{3}{2}\\\frac{5}{4}\end{array}\right.$$

$$20\{6\} + 12\{10\}$$

$$\sqrt{\frac{17 + 3\sqrt{5}}{2}}$$

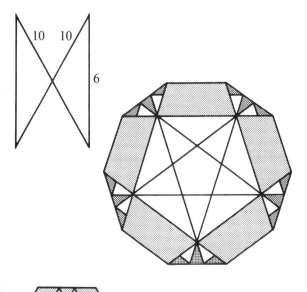

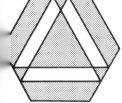

	1	2	3	4	5	6	7	8	9	10
(0)	Y	B	O	R	G	R	G	Y	B	O
(1)	W	Y	R	G	O	R	Y	O	B	G
(2)	(etc., in cyclic order down each column.)									

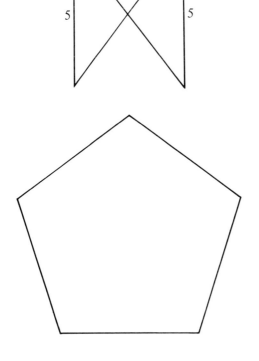

91 Small dodecahemi-dodecahedron

The relation of this polyhedron to **89** and to the icosidodecahedron has already been mentioned. The cups or inverted pyramids are trihedral holes with vertices at the centre of the polyhedron. The double tab or ribbed technique is the best to follow in making the model. Twenty triangular pyramids, identical to those of the great stellated dodecahedron **22** must be made with all double tabs turned outward to form ribs. The first ring of five pyramids can then be cemented together and a W pentagon added to serve as a base. This is actually the exterior surface of the completed solid. The pentagons are on planes parallel to the equatorial decagons and so should be the same colour. If you keep **89** in front of you as you work and remember that the triangles there are replaced here by the pentagons you should find it easy to complete the model. Again this makes a very rigid model.

$$\tfrac{5}{4}\ 5\,|\,5$$
$$12\{5\} + 6\{10\}$$
$$2\tau$$

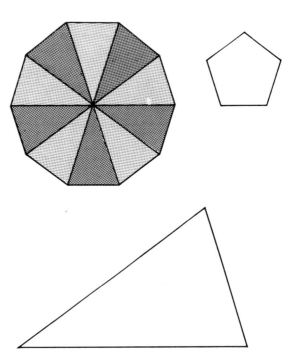

92 Quasitruncated hexahedron

This polyhedron is a quasitruncated cube. Six octagrams lie on the facial planes of an interior cube sharing their edges with the eight triangles whose planes intersect the cube. By using three colours for the octagram pairs and the other two colours for the triangles you can achieve a very suitable arrangement. To make a model of this polyhedron assemble the parts as shown.

Part I forms a cup with four pointed side faces and a square bottom. Part II forms a four-sided box, open at both ends, the bottom end straight and the top end jagged. Part I is cemented as a dimple into the jagged end of part II to form one section. Six of these sections complete the solid, the three described below in the colour table and their three enantiomorphs. You must be sure that part I is properly orientated before cementing it into part II, so the colours are correctly arranged. The sharp dihedral angles at these edges make cementing an easy process. This makes an attractive and rigid model.

$$2 \; 3 \mid \tfrac{4}{3} = t' \; \{4, 3\}$$
$$8\{3\} + 6\{\tfrac{8}{3}\}$$
$$\sqrt{(7 - 4\sqrt{2})}$$

Part I

1	2	3	4	5
O	R	G	R	G
B	R	G	R	G
Y	R	G	R	G

Part II

1	2	3	4
Y	B	Y	B
Y	O	Y	O
O	B	O	B

144

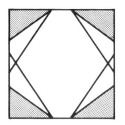

93 Quasitruncated cuboctahedron

This polyhedron has six octagrams in the same facial planes as a regular octahedron. The set of twelve squares intersect each other by threes in such a way that a set of eight small triangular holes provide openings that penetrate deeply into the interior of the solid. The sides and bottoms of these holes belong to the facial planes of the intersecting hexagons. Thus the sides of these holes are in fact a truncated version of the stella octangula and the bottoms coincide with the interior regular octahedron. This suggests the following method of assembly.

Begin by making the truncated pyramids shown, but turn all the tabs out, because this part is one of the holes. Since in an ordinary sized model the holes admit very little light, all the parts may as well be W. Double tabs are needed at the bases of part I, and are used to cement the parts together as in the regular octahedron. All these should be completed before proceeding to part II. Part II contains the central portion of the square planes, which can be given the colours Y, B, O, R and the two side wings which are W because they come from the hexagon planes. This part can then be cemented at its short edges to the edges of the small triangular holes of part I. Next, complete another set of these as colour pairs to the first four. A third set of four forms an equatorial band of squares. Finally pairs of triangles which belong to the corners of the square planes are cemented

$$2\ 3\ \tfrac{4}{3}| = t'\left\{\tfrac{3}{4}\right\}$$
$$8\{6\} + 12\{4\} + 6\{\tfrac{8}{3}\}$$
$$\sqrt{(13 - 6\sqrt{2})}$$

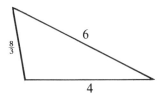

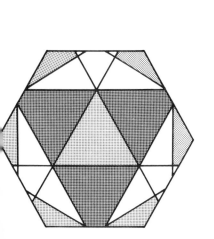

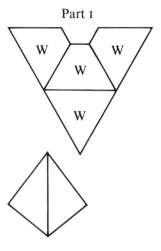

Part I

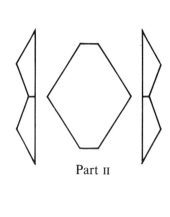

Part II

between alternate spaces around the octagram. If you watch the model as you complete it up to this point, it will not be difficult to determine the correct colours for these. This octagram, with its set of triangular grooves, can now very easily be cemented in place. The very acute overhang that occurs in this model makes it a simple matter to

handle, provided again that you finish an edge at a time. This makes a very sturdy model, because it already has an elaborate interior structure, but it also takes very careful work on this interior to get the exterior parts to fit well. It is interesting to note that the holes in this model generally go unnoticed by the inexperienced observer.

146

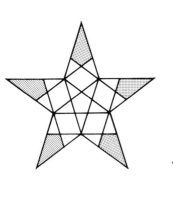

94 Great icosidodecahedron

This polyhedron is called the great icosidodecahedron because it has twenty triangles and twelve pentagrams in the same facial planes as the regular icosidodecahedron, yet it is not a stellation of the latter. A suitable colour scheme is to make all the pentagrams W and use the five colours for the triangles in the usual icosahedral order.

To construct a model of this polyhedron begin with the cups whose sides belong to five intersecting triangles and whose bottom is the small pentagon from the pentagram plane.

Next it is best to prepare the W star arms in sets of three, forming shallow trihedral dimples with the star arms pointing outwards as shown. These then serve as connectors between pentahedral dimples or cups. Enantiomorphism again applies to the second half of the model. This makes a very attractive model and proves to be very rigid.

$$2\,|\,3\ \tfrac{5}{2} = \begin{Bmatrix} 3 \\ \tfrac{5}{2} \end{Bmatrix}$$

$$20\{3\} + 12\{\tfrac{5}{2}\}$$

$$2\tau^{-1}$$

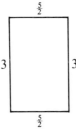

	0	1	2	3	4	5
(0)	W	Y	B	O	R	G
(1)	W	Y	B	G	O	R
(2)	W	B	O	Y	R	G
(3)	W	O	R	B	G	Y
(4)	W	R	G	O	Y	B
(5)	W	G	Y	R	B	O

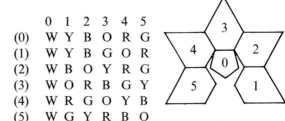

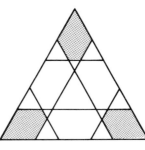

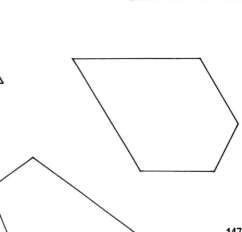

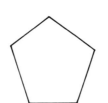

95 Truncated great icosahedron

This polyhedron is a truncated version of the great icosahedron. It may therefore follow the same colour arrangement. The hexagonal planes here take the place of the triangular planes. The method of assembly is therefore very similar. All the pentagrams may be W.

Begin with a W star and cement the parts shown below between the star arms. Follow the same paired arrangement as in the colour table for the great icosahedron **41**. When the isosceles triangles have been cemented you will have completed one section of the model, a pentahedral dimple serving as a sort of tray for a raised star. The next five sections are done in the same way, each with a W star. These sections are then cemented to each other in the same way as the vertex parts of the great icosahedron. Equally you may choose to do all the stars in the dodecahedral arrangement using opposite stars of the same colour and requiring six colours in all. But then you cannot avoid having each star share one of its edges with a hexagon of the same colour. However this does not detract from the beauty of this model, because the two planes make a sharp angle, almost a 90° turn, at these edges.

$$2 \tfrac{5}{2} | 3 = t \{3, \tfrac{5}{2}\}$$
$$20\{6\} + 12\{\tfrac{5}{2}\}$$
$$\sqrt{\frac{29 - 9\sqrt{5}}{2}}$$

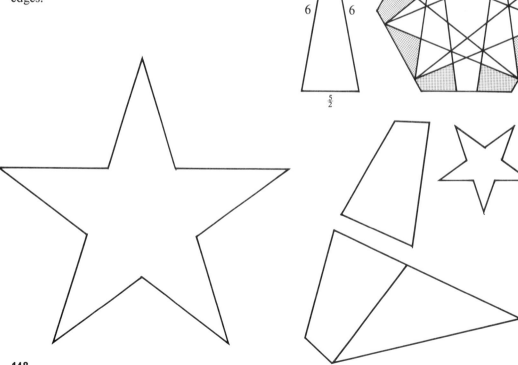

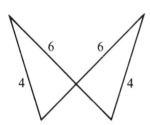

96 Rhombicosahedron

This polyhedron is closely related to **76** and **83**. They all have the same set of thirty intersecting skew squares forming equatorial bands. The hexagons take the place of the pentagons and thus the star planes become faceted stars. Also a set of thirty shallow cups appear here directly over the middle section of each square, the bottom of each cup being part of the square plane, and its four sides, each deeply recessed, being part of four hexagon planes. It would be more tedious than necessary to make a model of this polyhedron while maintaining the same colours for each plane. The method of assembly suggested here will maintain different colours for the squares and the upper surface of the hexagon planes. This will still give a very attractive result and only an informed observer would notice the colour discrepancies since they occur only on the deeply overhanging underside of the hexagons and thus are hardly visible. You may find even the suggested method of assembly tedious enough because of the large number of parts involved, so again your perseverance will be put to the test if you wish to complete the entire model.

Begin by assembling a faceted star as shown. Each of these has a central cup made of five equilateral triangles in the first icosahedral arrangement. The isosceles triangles in the star arms are next cemented to the edges of the central cup. These are parts of the square planes. A colour table for six parts follows.

$$2\ 3\ \frac{\frac{5}{4}}{\frac{5}{2}}$$

$$20\{6\} + 30\{4\}$$

$$\sqrt{7}$$

	1	2	3	4	5	6	7	8	9	10
(0)	Y	B	O	R	G	R	G	Y	B	O
(1)	Y	R	O	G	B	G	O	W	Y	R
(2)	B	G	R	Y	O	Y	R	W	B	G
(3)	O	Y	G	B	R	B	G	W	O	Y
(4)	R	B	Y	O	G	O	Y	W	R	B
(5)	G	O	B	R	Y	R	B	W	G	O

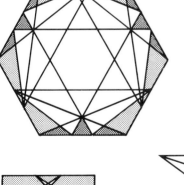

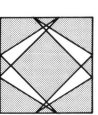

Part I

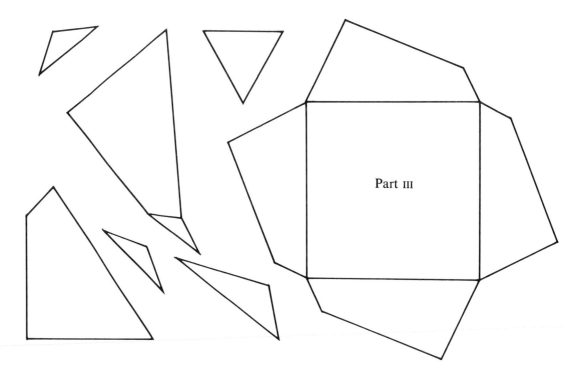

Part III

When you have assembled these ten parts of a faceted star, you will have no difficulty cementing the pairs of obtuse triangles that finish the dimples in the star arms. These pairs lie in the planes of the equilateral triangles in the central cup and so the colours are determined by following the plane out to the star arm. Again these faceted stars lack rigidity, so it is better to add some backing. The pairs shown below, as part II, are cemented between the star arms. They have the same colour arrangement as in **76**, but here the shape is different. You will notice a small triangle at the blunted vertex of this part. It is a tiny portion of the underside of the hexagon planes, so the colour will not have to be the same as the upper side of these planes. Assembly is greatly simplified if it is cut in one piece with the quadrilateral to which it is attached. Part III is also best done as one net. You will need thirty of

these, five of each of the colours W, Y, B, O, R, G. Part III forms a shallow cup with deeply recessed sides which are almost invisible when viewed straight on. Cement five of these cups between the pairs in part II. The colour sequence is determined by the corners of the squares formed by part II.

Continue making faceted stars, each with a set of five of part II between the star arms and cement them together using the cups of part III as connectors. Finally the small triangular holes left at the blunt ends of three of part II are closed last of all with the small trihedral dimples shown in part IV, the faces of which belong to the square planes. The dimples are best cut without tabs as they will then easily fit into their positions. The colour arrangement is followed by watching the order of colours in the intersecting squares. This makes a fairly rigid model and is certainly noteworthy for its complexity.

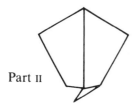

Part II

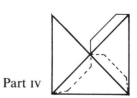

Part IV

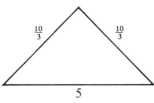

97 Quasitruncated small stellated dodecahedron

This polyhedron is a quasitruncated version of the small stellated dodecahedron. A model can easily be assembled in twelve sections whose base edges are equivalent to the edges of the dodecahedron. The colour arrangement is the usual six-colour dodecahedral one, set out in the table below.

The top edges of 6, 7, 8, 9, 10 are cemented to 1, 2, 3, 4, 5. The section then has a pentagonal edge at the bottom. The tabs at these edges are cemented to each other as if the sections were faces of a dodecahedron. Notice that 1 is on a plane parallel to 6, 2 to 7, and so on, making decagram and pentagram planes the same colour. This makes a very attractive and rigid model.

$$2\ 5\left|\tfrac{5}{3} = t'\left\{\tfrac{5}{2}, 5\right\}\right.$$
$$12\{5\} + 12\{\tfrac{10}{3}\}$$
$$\sqrt{\frac{17 - 5\sqrt{5}}{2}}$$

	0	1	2	3	4	5	6	7	8	9	10
(0)	W	Y	B	O	R	G	Y	B	O	R	G
(1)	Y	W	G	O	R	B	W	G	O	R	B
(2)	B	W	Y	R	G	O	W	Y	R	G	O
(3)	O	W	B	G	Y	R	W	B	G	Y	R
(4)	R	W	O	Y	B	B	W	O	Y	B	G
(5)	G	W	R	B	O	Y	W	R	B	O	Y

98 Quasitruncated dodecahedron

This polyhedron is to the great stellated dodecahedron what **88** is to the stellated octahedron. Here the decagrams replace the octagrams. A set of thirty squares intersecting by threes appears here, just as twelve squares appeared in **88**, forming twenty small triangular holes which penetrate deeply into the interior. The facial planes of these holes belong to the intersecting decagons.

The simplest way to make this model is to use only three colours, one for each type of polygon. Begin as in **88** where you built the interior truncated version of the stellated octahedron. Here you build the interior truncated version of the great stellated dodecahedron to form the sides of the holes, but the bottoms are equivalent to the surface of the great dodecahedron. Since this will all be one colour, one net serves for all twenty parts (part I). Remember to turn the tabs out so these parts can be cemented together. All twenty are best completed before continuing with the rest of the model.

Next, prepare the central parts of the squares with their adjoining wings, the squares of a second colour, the wings of the same colour as part I. The figure shows the arrangement of nets for part II. Part II is then cemented to straddle from hole to hole. Next the decagrams are prepared with the triangle pairs from the square plane corners cemented between every other star arm. This is analogous to the procedure followed in **88**. The deep recess of part II makes it easy to cement these decagram parts, although the large number of edges will take some time. With perseverance you will finish a sturdy and pleasing model.

$$2 \tfrac{5}{3} \, 5 \,|= t' \begin{Bmatrix} \tfrac{5}{2} \\ 5 \end{Bmatrix}$$
$$12\{10\} + 30\{4\} + 12\{\tfrac{10}{3}\}$$
$$\sqrt{11}$$

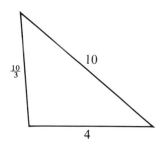

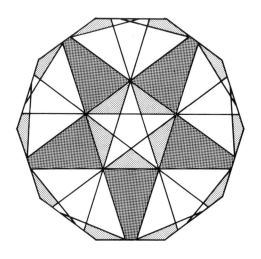

Part I

Part II

99 Great dodecicosido-decahedron

This polyhedron is very attractive when done in colour because it readily appears as a set of highly embossed and richly designed solid stars on decagram planes. It is easy to make this model by the usual sectional method of assembly. Part I is shown with its chevron-shaped faces. These form trihedral dimples with the same colour arrangement as the great dodecahedron **21**, so you may refer back to it for your requirements here. Part II is made up of star arm pairs and these serve as connectors for part I. Each of these is cemented so the star arm colours are on planes above and parallel to the same colour of the chevron-shaped faces of part I. The five rhombi from the triangle planes, part III, form a rosette. Part III follows the first icosahedral arrangement of colours. As soon as you have cemented a ring of five of each of part I and part II alternating, the first rosette can be cemented to fill the hole at the centre. Once this has been done the colours for the other pairs of part II are easy to determine because the parallel planes take the same colour. In cementing the rosettes keep an eye on the triangular planes.

$$3 \ \tfrac{5}{2} \,|\, \tfrac{5}{3}$$
$$20\{3\} + 12\{\tfrac{5}{2}\} + 12\{\tfrac{10}{3}\}$$
$$\sqrt{(11 - 4\sqrt{5})}$$

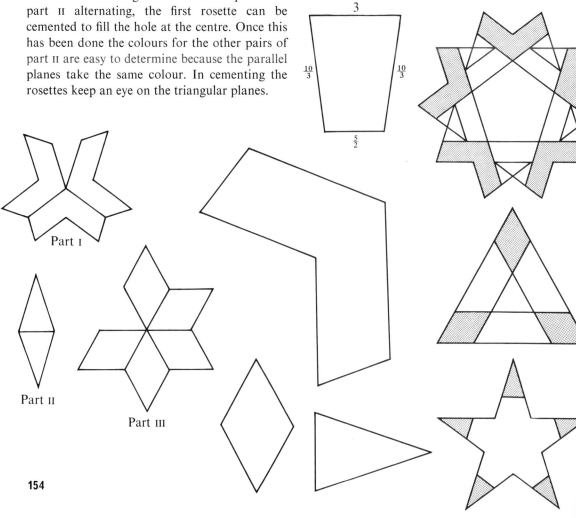

Part I

Part II

Part III

154

100 Small dodecahemi-cosahedron

This polyhedron has twelve pentagrams on the facial planes of a dodecahedron and ten equatorial hexagons whose centre points all coincide with the centre of the solid. It is related to **73** whose dimples are here replaced with deep hexahedral holes or inverted pyramids. These holes may be thought of as the twelve vertex parts of the great icosahedron, all inverted and turned inwards. The colour arrangement suggested below will show very effectively an important feature of polyhedral symmetry, namely the generation of spherical lunes.

Turn all the tabs outward to form ribs and then use these to cement the parts to each other, the tabs of one under the surface of the other so the edges coincide. W pentagrams are cemented as needed. The usual enantiomorphism applies, and you should have no difficulty completing the model. This makes a very rigid model, but care must be exercised to make the pentagrams accurately so they fit well. The final result also depends on the accuracy of the inverted pyramids.

$$\tfrac{5}{3}\ \tfrac{5}{2}\,|\,3$$
$$10\{6\} + 12\{\tfrac{5}{2}\}$$
$$2$$

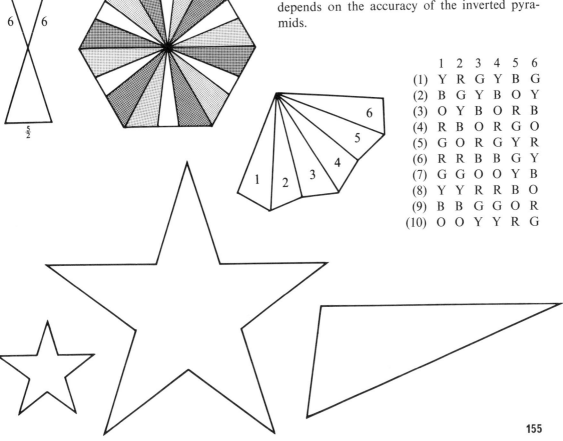

	1	2	3	4	5	6
(1)	Y	R	G	Y	B	G
(2)	B	G	Y	B	O	Y
(3)	O	Y	B	O	R	B
(4)	R	B	O	R	G	O
(5)	G	O	R	G	Y	R
(6)	R	R	B	B	G	Y
(7)	G	G	O	O	Y	B
(8)	Y	Y	R	R	B	O
(9)	B	B	G	G	O	R
(10)	O	O	Y	Y	R	G

101 Great dodecicosahedron

This polyhedron is like **81** except that the dimples and grooves there are here replaced with deeper holes and cups, nonahedral and tetrahedral. The usual dodecahedral and icosahedral arrangement of colours applies again. The model can be assembled by making the parts as shown. These alternate between the star arms and serve as connectors.

You can easily see which position each part II should take if you keep the intersecting hexagons, each its own colour. You will also find the cups sharing edges internally but the tabs on these edges need only be adjusted, not cemented.

$$3 \; \tfrac{5}{3} \; \begin{vmatrix} \tfrac{3}{2} \\ \tfrac{5}{2} \end{vmatrix}$$

$$20\{6\} + 12\{\tfrac{10}{3}\}$$

$$\sqrt{\dfrac{17 - 3\sqrt{5}}{2}}$$

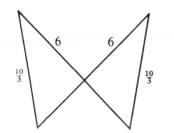

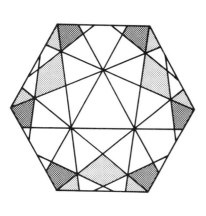

Part I

	1	2	3	4	5	6	7	8	9
(1)	B	Y	R	B	B	G	Y	G	Y
(2)	O	B	G	O	O	Y	B	Y	B
(3)	R	O	Y	R	R	B	O	B	O
(4)	G	R	B	G	G	O	R	O	R
(5)	Y	G	O	Y	Y	R	G	R	G
(6)	B	O	Y	B	G	O	R	B	R
(7)	O	R	B	O	Y	R	G	O	G
(8)	R	G	O	R	B	G	Y	R	Y
(9)	G	Y	R	G	O	Y	B	G	B
(10)	Y	B	G	Y	R	B	O	Y	O

Part II

	1	2	3	4
(1)	Y	G	Y	R
(2)	B	Y	B	G
(3)	O	B	O	Y
(4)	R	O	R	B
(5)	G	R	G	O
(6)	Y	G	Y	O
(7)	B	Y	B	R
(8)	O	B	O	G
(9)	R	O	R	Y
(10)	G	R	G	B
(11)	B	G	O	Y
(12)	O	Y	R	B
(13)	R	B	G	O
(14)	G	O	Y	R
(15)	Y	R	B	G

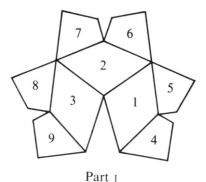

Part I

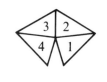

Part II

156

157

102 Great dodecahemicosahedron

This polyhedron is a faceted version of the dodecadodecahedron. The facial planes of the faceted stars are a combination of the intersecting pentagons and hexagons, whose parts are easily recognizable. The hexagons are in the same planes as those of **100**. To make a model of this polyhedron make the faceted stars using the colour table given below.

The figure is merely a plan and does not show the exact shape of each part, only its relative position. The 0 and 6 7 8 9 10 parts belong to the pentagon, the rest to the hexagon. These faceted stars are then joined by using the dimples between star arms as in **73**. This makes a very attractive model and is very rigid without any further internal support.

$\frac{5}{4}$ 5|3

$10\{6\} + 12\{5\}$

2

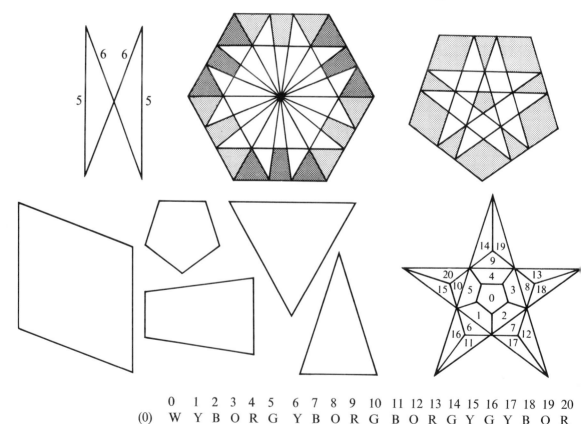

	0	1	2	3	4	5	6	7	8	9	10	11	12	13	14	15	16	17	18	19	20
(0)	W	Y	B	O	R	G	Y	B	O	R	G	B	O	R	G	Y	G	Y	B	O	R
(1)	Y	Y	B	G	B	G	W	G	O	R	B	B	G	B	G	Y	G	Y	B	G	Y
(2)	B	B	O	Y	O	Y	W	Y	R	G	O	O	Y	O	Y	B	Y	B	O	Y	B
(3)	O	O	R	B	R	B	W	B	G	Y	R	R	B	R	B	O	B	O	R	B	O
(4)	R	R	G	O	G	O	W	O	Y	B	G	G	O	G	O	R	O	R	G	O	R
(5)	G	G	Y	R	Y	R	W	R	B	O	Y	Y	R	Y	R	G	R	G	Y	R	G

103 Great rhombihexahedron

This polyhedron is closely related to **77**. The tetrahedral dimples and dihedral grooves of that solid are here replaced with deeper nonahedral cups and tetrahedral dimples. Three colours can again be used for the octagrams, making opposite pairs the same colour. All the other planes are squares intersecting internally. Since there are twelve of these, six colours will serve as six pairs. This means of course that the Y squares must meet the Y stars along at least one edge, and so on for each of the other colours, but the sharp angle at these edges helps greatly to make this violation of the map colouring principle almost unnoticable.

The best method of construction is to make the cups and dimples first, then to cement them alternately between the star arms. You will find

$$2 \ \frac{4}{3} \ \begin{vmatrix} \frac{3}{2} \\ \frac{4}{2} \end{vmatrix}$$

$$12\{4\} + 6\{\tfrac{8}{3}\}$$

$$\sqrt{(5 - 2\sqrt{2})}$$

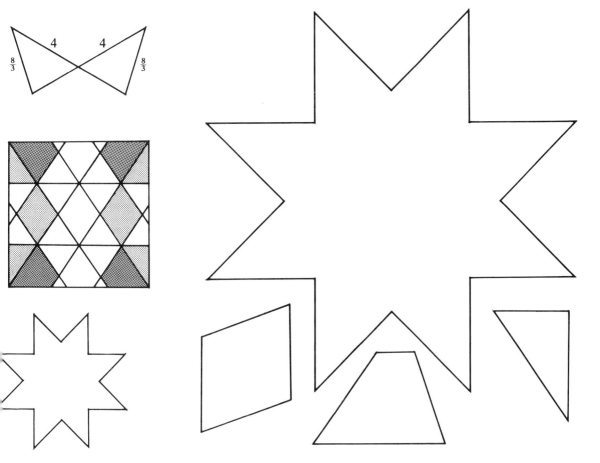

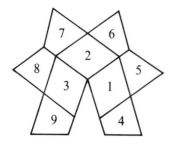

Part I

	1	2	3	4	5	6	7	8	9
(1)	Y	O	B	G	W	R	G	W	R
(2)	B	W	R	Y	O	G	Y	O	G
(3)	R	O	G	B	W	Y	B	W	Y
(4)	G	W	Y	R	O	B	R	O	B

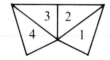

Part II

	1	2	3	4
(1)	Y	O	W	R
(2)	B	W	O	G
(3)	R	O	W	Y
(4)	G	W	O	B

that adjoining cups and dimples share common edges internally but the double tabs at these edges need only be adjusted, not cemented.

If you watch the parts of the square planes to see that each keeps its own colour you should have no trouble finding the correct position for each part. All the parts listed above are cemented to a Y octagram. Once this is done you are ready to cement the next four octagrams, O, B, O, B. These will then be completely surrounded by cups and dimples whose colour arrangement is enantiomorphous to the first set and diagonally opposite their counterparts on the completed model. Another Y octagram is added last, one edge cemented at a time in the usual manner. The deeper cups and dimples make this model even more attractive than **77**.

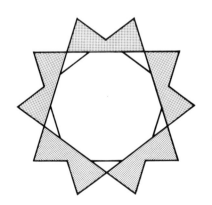

104 Quasitruncated great stellated dodecahedron

This polyhedron is a quasitruncated version of the great stellated dodecahedron. The triangle planes cut the vertices of the latter almost at their bases, so that the pentagrams are transformed into decagrams. Trihedral dimples close the cut ends. Begin this polyhedron by making the truncated pyramids, one of whose side faces is shown below. The colour arrangement for these is the same as that for the great stellated dodecahedron **22**, so you may simply refer back to it for the colour table to be used here. The trihedral dimples have the icosahedral arrangement of colours, as given in the table.

You must watch carefully how you orient these dimples before cementing them into their positions closing the jagged end of the truncated pyramids. (1) must have its Y arm cemented between the B and G of the pyramid, and so on round. Once you have completed the first ring of truncated pyramids and their dimples, the other parts are easier to locate. Since opposite dimples are not enantiomorphs the complete colour table is given.

$$2\ 3\,|\,\tfrac{5}{3} + t'\,\{\tfrac{5}{2}, 3\}$$
$$20\{3\} + 12\{\tfrac{10}{3}\}$$
$$\sqrt{\dfrac{37 - 15\sqrt{5}}{2}}$$

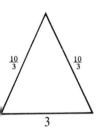

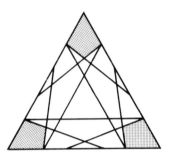

	1	2	3		1	2	3
(1)	Y	G	O	(11)	B	Y	G
(2)	B	Y	R	(12)	O	B	Y
(3)	O	B	G	(13)	R	O	B
(4)	R	O	Y	(14)	G	R	O
(5)	G	R	B	(15)	Y	G	R

	4	5	6		4	5	6
(6)	R	G	Y	(16)	Y	B	R
(7)	G	Y	B	(17)	B	O	G
(8)	Y	B	O	(18)	O	R	Y
(9)	B	O	R	(19)	R	G	B
(10)	O	R	G	(20)	G	Y	O

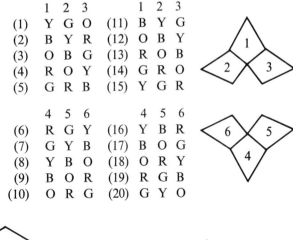

105 Quasirhombicosidodecahedron

This polyhedron is very similar to **99**, except that here the decagram planes give way to a fantastically intricate network of intersecting planes generating a very interestingly faceted structure. Fortunately it is not too difficult to make and only requires plenty of perseverance to cut and join the numerous parts. The arrangement of colours for triangles and pentagrams used in **99** can be retained here, but the squares are best in white. This vastly simplifies the assembly.

Begin with a dimpled rosette of five colours as in **99**. Surround this with five parts, all W and of one net as shown for part I. Cement a set of five star arms to the side parts. The Y star arm shares its vertex with the Y rhombus of the rosette, and so on for the other colours. Then make a set of paired wedges, as shown for part II, whose triangular faces are very small parts of the square and triangle faces. The tabs on the shortest sides of these triangles are cemented to the tabs at the small cut-out of part I. The best way to handle these wedges is to cement the W first, one tab at a time, then both coloured tabs at once using tweezers, since the parts are so small in a normal sized model. The last parts to be cemented for this section can now be done, one edge at a time. These are the ten pieces, shown as part IA, belonging to one facial plane of the pentagram. Once these are added you should have a fairly rigid structure whose projection on the plane of the pentagram is shown opposite. This completes one section of the model. A total of twelve are required altogether. The first or (0) section has the first set of five sections surrounding it. These are cemented by one of

$$3 \; \tfrac{5}{3} \,|\, 2 = r'\left\{\begin{matrix}3\\ \tfrac{5}{2}\end{matrix}\right\}$$
$$20\{3\} + 30\{4\} + 12\{\tfrac{5}{2}\}$$
$$\sqrt{(11 - 4\sqrt{5})}$$

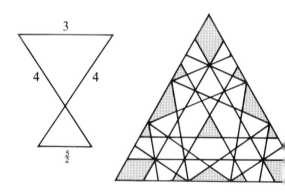

$$\tau = \tfrac{1}{2}(\sqrt{5}+1) = 1\cdot618$$
$$\tau^{-1} = \tfrac{1}{2}(\sqrt{5}-1) = 0\cdot618$$
$$\tau^{-5} = 0\cdot09$$

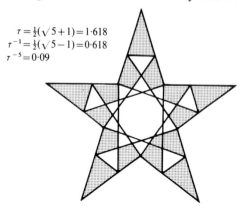

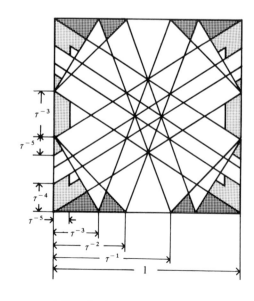

162

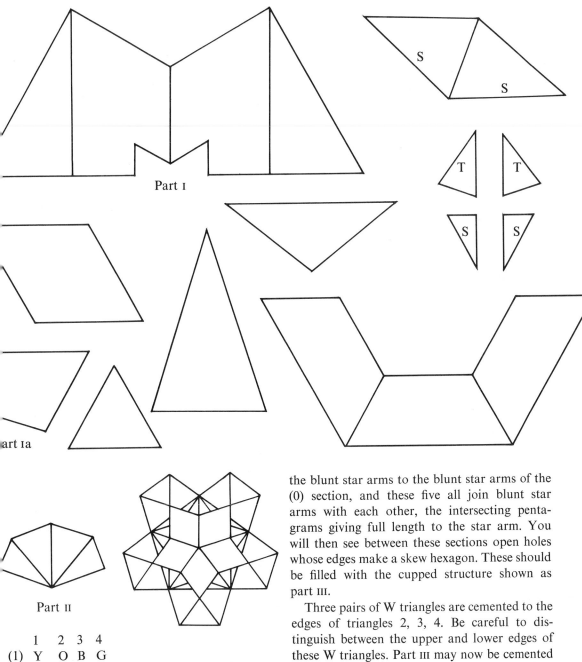

Part I

Part Ia

Part II

Part III

	1	2	3	4
(1)	Y	O	B	G
(2)	B	R	O	Y
(3)	O	G	R	B
(4)	R	Y	G	O
(5)	G	B	Y	R
(6)	Y	R	B	O
(7)	B	G	O	R
(8)	O	Y	R	G
(9)	R	B	G	Y
(10)	G	O	Y	B

The rest are the enantiomorphs.

the blunt star arms to the blunt star arms of the (0) section, and these five all join blunt star arms with each other, the intersecting pentagrams giving full length to the star arm. You will then see between these sections open holes whose edges make a skew hexagon. These should be filled with the cupped structure shown as part III.

Three pairs of W triangles are cemented to the edges of triangles 2, 3, 4. Be careful to distinguish between the upper and lower edges of these W triangles. Part III may now be cemented to fill the skew hexagonal holes. If you watch the triangles 2, 3, 4 so that their colours agree with those in the rosette you can get part III correctly placed. The triangle 1 of part III belongs to the central portion of the triangle plane.

This turns out to be a very beautiful model, so it is well worth the time it takes to assemble it. You may expect to spend at least 30 hours to complete it.

106 Great icosihemi-dodecahedron

This polyhedron is closely related to **94**. The pentahedral dimples or rosettes formed by the triangle planes are still here, but the pentagrams have disappeared giving way to equatorial deca-grams whose facial planes form deep cups where the star arms used to be. This suggests a method of assembly. Make the cups first, each with double tabs turned out to form ribs in the usual manner. Use these ribs as tabs a second time to cement a set of five cups in a ring. The colour arrangement for these follows:

	1	2	3	4
(1)	Y	G	Y	B
(2)	B	Y	B	O
(3)	O	B	O	R
(4)	R	O	R	G
(5)	G	R	G	Y
(6)	Y	O	W	G
(7)	B	R	W	Y
(8)	O	G	W	B
(9)	R	Y	W	O
(10)	G	B	W	R

The rest are the enantiomorphs.

The same icosahedral arrangement as in **94** is also used here for the rosettes. These may be cemented in place as each ring of cups is completed. Keep your attention on the facial planes, both for the cups and the rosettes to see that each decagram plane and each triangle plane keeps its own colour. To help get the first rosette properly oriented, place the Y rhombus so that it matches the small Y triangle at the bottom of one of the cups in the first ring of five. The rest then follow in order.

$$\tfrac{3}{2}\ 3\,\Big|\,\tfrac{5}{3}$$
$$20\{3\} + 6\{\tfrac{10}{3}\}$$
$$2\tau^{-1}$$

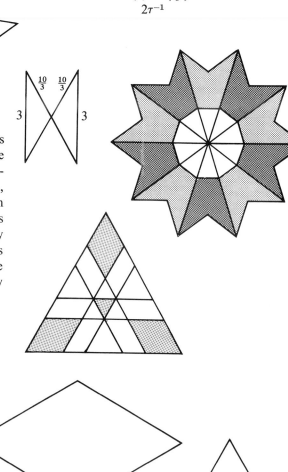

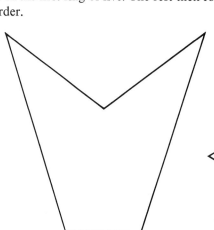

164

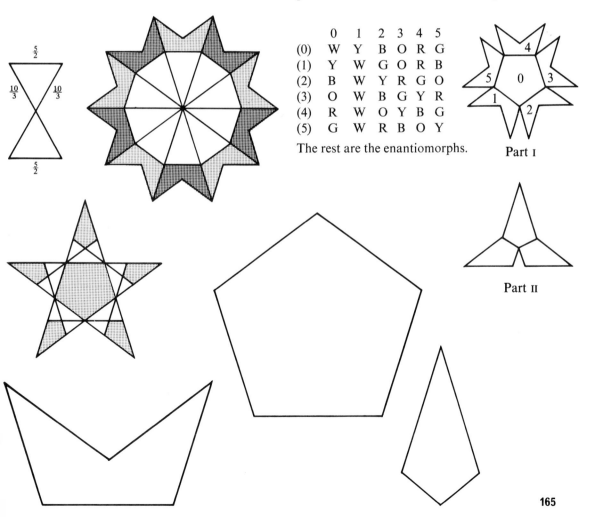

107 Great dodecahemi-dodecahedron

This polyhedron is closely related to **94** and to **106**. It is simple and very rigid in its structure, with equatorial decagrams on planes parallel to and midway between pairs of opposite pentagrams. This is beautifully brought out in the colour arrangement. To make a model of this polyhedron begin with the pentagonal cups as listed in the colour table below.

Cement each cup of part I using the double tabs as connecting tabs, turned slightly and brought under the surface of its neighbour, to make the slant edges of the cups coincide. Then make twenty star arm dimples, the trihedral arrangement being the same as the colour table used for the great dodecahedron. These are used to close the holes between the cups, matching the same planes for colour. This model is very attractive.

$$\frac{5}{3} \ \frac{5}{2} \ | \ \frac{5}{3}$$
$$12\{\tfrac{5}{2}\} + 6\{\tfrac{10}{3}\}$$
$$2\tau^{-1}$$

	0	1	2	3	4	5
(0)	W	Y	B	O	R	G
(1)	Y	W	G	O	R	B
(2)	B	W	Y	R	G	O
(3)	O	W	B	G	Y	R
(4)	R	W	O	Y	B	G
(5)	G	W	R	B	O	Y

The rest are the enantiomorphs.

Part I

Part II

165

108 Great quasitruncated icosidodecahedron

This polyhedron has very many parts, and in a model of ordinary scale the faceted star parts are very small. For this reason a full colour model would entail a very tedious amount of work. The method of assembly described here limits the colours to three, one for each of the three types of polygons that appear as facial planes. A good colour combination is obtained by using Y for the hexagons, R for the squares, and B for the decagons. Begin by making a set of five parts of each of the pairs shown, parts I (a) and (b). These ten parts are cemented alternately in a ring, leaving a star shaped hole in the centre. This hole is now filled with a tiny faceted star or pentagram. They are shown full scale for a model of 8 inches on an edge. Make the central pentahedral dimple first. Then cement the small triangles to its edges. Next add the pairs of triangles completing the star arm dimples and finally add the other paired triangles between the star arms. This dimpled or faceted star can now be cemented to close the star shaped hole left in the ring of part I parts. You will have to make a total of twelve rings and twelve faceted stars for the complete model. These twelve sections are joined with three types of connectors, a nonahedral cup, of six side faces and three bottom faces (part III a), a dihedral groove (part III b), and a tetrahedral dimple (part III c). These are shown in full scale as part III a, b, c. These connecting parts help to give some rigidity to the model, but for best results some further interior supports are needed, especially under the edges of part I.

$$2 \ 3 \ \tfrac{5}{3} \,\big| = t' \begin{Bmatrix} 3 \\ \tfrac{5}{2} \end{Bmatrix}$$
$$20\{6\} + 30\{4\} + 12\{\tfrac{10}{3}\}$$
$$\sqrt{(31 - 12\sqrt{5})}$$

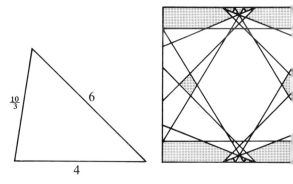

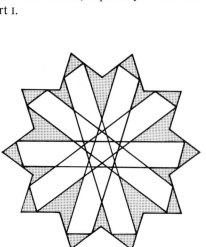

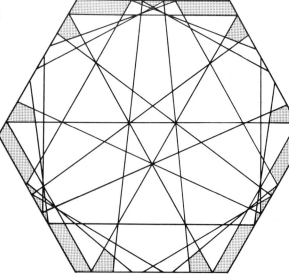

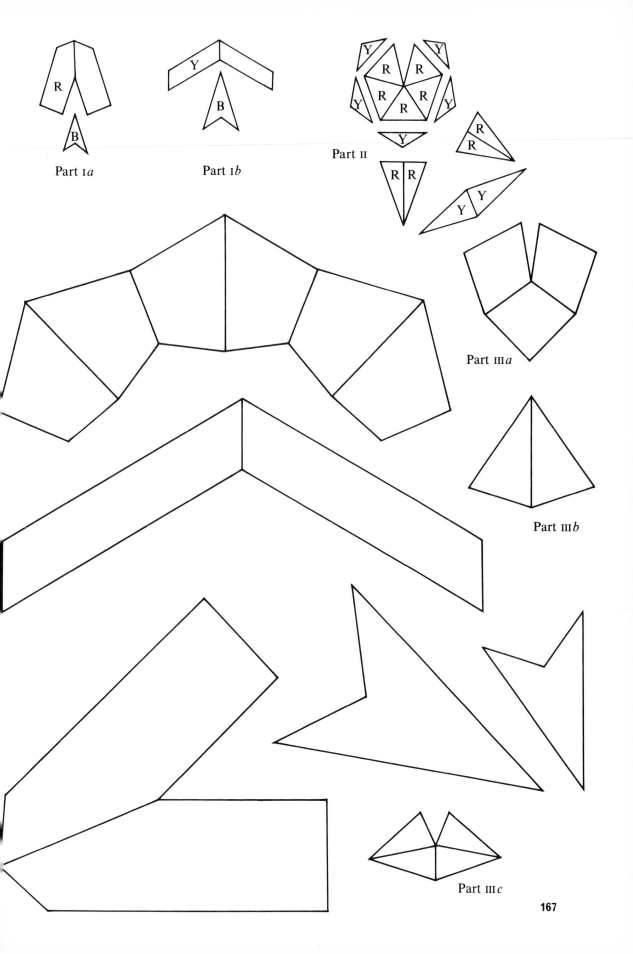

Part Ia

Part Ib

Part II

Part IIIa

Part IIIb

Part IIIc

167

109 Great rhombidodecahedron

This polyhedron has the same decagrams as **99**, but the pentagrams and triangles there are replaced by squares here. The intersection of these squares with one another introduces deep holes at the places where the twin star arms and triangle rosettes occurred in **99**. In place of the twin star arms you will find hexahedral cups whose four sides belong to the facial planes of four squares and whose bottom is a pair of small isosceles triangles of the 36°, 72° variety belonging to the facial planes of two decagrams. In place of the rosettes you will find a fascinatingly complex structure of cavities which fortunately is not too difficult to construct. As for colour arrangements you undoubtedly will agree that it would be much too tedious to use six colours for the squares. So the suggestion here is to make all the squares W and then distribute the six colours Y, B, O, R, G, and W according to the usual dodecahedral arrangement as in **99**. In this way the chevron-shaped parts are used once more here. Thus you may begin the construction of this model with a set of five trihedral chevron-shaped dimples as in **99**. These are part I. The other parts are also set out full scale for a model whose edges will be 8 inches.

Part II shows the hexahedral cup, the sides being one net for parts all W. The small triangles must be done in colour pairs determined by their position relative to part I, so you should need no specific directions to get them right if you work systematically. Since they are so small it is easiest to cement them as pairs, then each pair is cemented as a bottom to the W cups. Five of part I and five of part II then form a ring with a hole in the centre. This hole is in the shape of a skew decagon. Next you must make five of part III as shown. All have the paired triangles in W as one net with the other part from the decagram planes in the usual dodecahedral order of colours. When completed part III forms a trihedral cup pointed at the bottom, the upper edges having the shape of a skew quadrilateral. These parts are then immediately cemented so

$$2 \ \frac{5}{3} \ \frac{\frac{3}{2}}{\frac{5}{4}} \Big|$$

$$30\{4\} + 12\{\tfrac{10}{3}\}$$

$$\sqrt{(11 - 4\sqrt{5})}$$

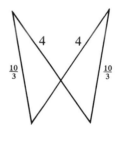

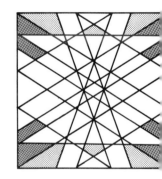

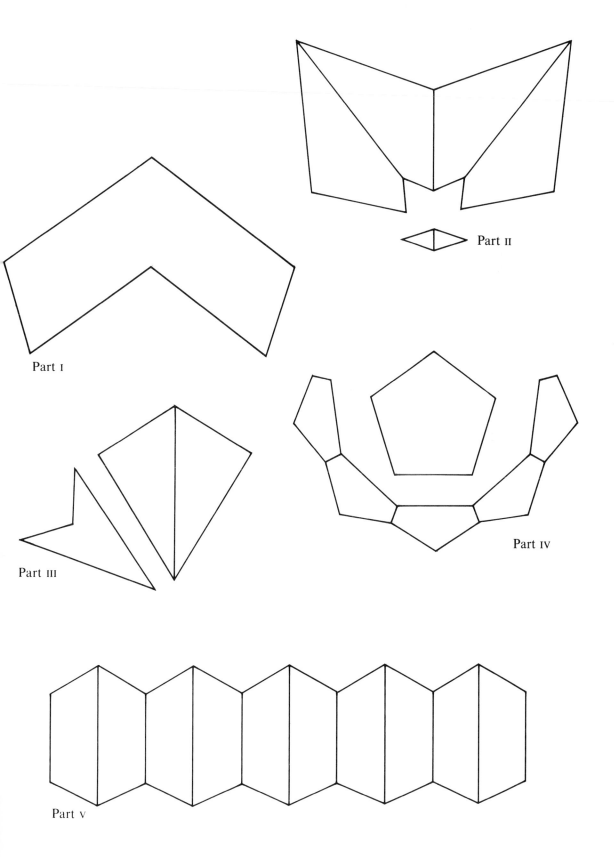

Part I

Part II

Part III

Part IV

Part V

the angle formed by the W edges is vertically opposite the angle formed by the two sides of the part II cups. You will now see the decagram planes being filled in, so you can get the parts placed correctly by watching this. When these parts have all been cemented you will still find a hole in the centre, still a skew decagon but deeper down. This hole is now again filled with another cup shown as part IV. A net for the sides, all W, is shown p. 169, part V. You must be careful to distinguish between the upper and lower ends of the series of ten trapezia, because they are not the same, but very nearly so. The series makes a ring in the form of a ten sided tube or prism. It is all W since all these trapezia belong to the facial planes of ten intersecting squares. The ring of irregular pentagons shown as part IV is also one net all W since they belong to the facial planes of five more squares. You can see now why it would be tedious to make all of these square planes a different colour. The ring of irregular pentagons may now be cemented to the lower end of the ten sided prism. The regular pentagon then forms the bottom and closes off that end. It

takes the colours of the decagram planes since it is the central portion of those planes. The tabs on this multifaced cup are of course all turned outward as ribs on the outside of the cup, which is then gently forced down the remaining central hole left by the five parts of part III and cemented an edge at a time. The reason for exerting 'gentle force' is that this cup has vertical sides which just fit into the hole so you must adjust the outer ribs while you are placing it in its position. Clamps are helpful in cementing twenty sets of tabs, and with a little skill and perseverance you should succeed.

You may actually complete the whole model as a shell without filling any of the central holes. This has the advantage of giving you the overall colour scheme, especially for parts II and IV. However, complete rigidity is not achieved until all the central holes are closed. Once this is done you have a very attractive model, very interesting because of its intricate structure. Since all the squares are W, their facial planes are not too evident, but with the model in your hands they are not hard to locate.

Commentary on non-convex snub polyhedra

There are two convex snub polyhedra, the snub cube and the snub dodecahedron. Among the non-convex polyhedra there are at least nine, ten if you count **119**, which is rather different from the others. The snub quality manifests itself in the two convex cases, in the twisted way in which the squares and pentagons are related to a circumscribing cube and dodecahedron respectively. The twist introduces *dextro* and *laevo*, right- and left-handed varieties in each case, and also a special set of triangular faces, the snub triangles. The same twisted characteristic is found in the non-convex cases and also sets of snub triangles. Model **119** is special in that the diametral squares may be considered as the snub faces.

The patterns on the facial planes of these non-convex snubs are tantalizingly irregular. None of the usual symmetry manifests itself, except, surprisingly enough, in **110** which is very simple and in **118** which is very complex. Because of this lack of facial symmetry (the solids as such have rotational symmetry), the intersections of the facial planes determining the pattern on these faces can only be found by calculation. This involves the use of analytic or coordinate geometry. The use of the usual Euclidean tools, ruler and compasses alone, will not suffice

Mr R. Buckley of Windsor, Berks, England, has recently performed these calculations, obtaining all the numerical data by programming a computer to work out the analytic equations. Briefly stated, his method involves a system of spherical coordinates, the polar axis being along an axis of rotational symmetry and the zero meridian through a vertex of the polyhedron. Then with right and left specified, polar coordinates for all vertices can be calculated. These are translated into the usual Cartesian coordinates to obtain equations for the facial planes. Then the line of intersection of a pair of facial planes is readily obtained by solution of a system of equations. The computer supplied numerical results correct to six significant figures. For the purposes of model-making this degree of accuracy is unnecessary.

The drawings given here were derived from large-scale drawings, edge length 20 cm (for **117** and **118**, 20 inches) supplied by Mr Buckley. Only some of the principal lines of intersection are given. The exterior portions of each facial plane are shaded as usual, light grey for the upper side and dark grey for the under-side. Numerical data has been reduced to two figures. All the models shown in the photographs are less than 12 inches in height, except for **117** and **118** which are nearly 24 inches tall for an edge length of 20 inches. The nets included in the instructions will give you models on this scale.

110 Small snub icosicosi-dodecahedron

This is the first of the non-convex snub polyhedra, and it is the simplest to make. Twenty pairs of equilateral triangles, forty in all, share icosahedral facial planes, giving this polyhedron the appearance of having twenty hexagram faces. But these hexagrams are not quite regular, although the triangles from which they are formed are equilateral, as they must be in a uniform polyhedron. The twelve pentagrams are completely surrounded by another set of triangles, sixty in all. This suggests a simple method of construction.

Begin by cementing the scalene triangular parts as dihedral grooves between pentagram star arms, following the icosahedral arrangement of colours. The pentagrams are all the same colour, W. This is part I of the model. The first section is shown below, the (0) section. Five hexagrams (part II) are then cemented around part I. These may be a seventh colour or the usual Y, B, O, R, G arranged to maintain the map-colouring principle. The next five sections of part I are then cemented in place. You will now see openings between the hexagrams. Close these with pairs of small equilateral triangles, part III, each the colour appropriate to the facial plane to which it belongs. Cement the pairs together first and then cement them to close the openings between the hexagrams.

$$|3\ 3\ \tfrac{5}{2}$$
$$(40+60)\{3\}+12\{\tfrac{5}{2}\}$$
$$2{\cdot}91638\ 06615$$

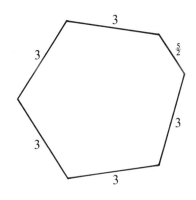

Edge length = 6·6 cm
$a = 2{\cdot}1$
$b = 2{\cdot}3$
$c = 2{\cdot}5$

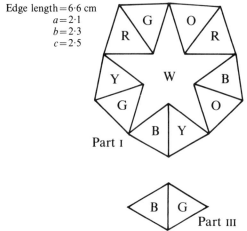

Part I

Part III

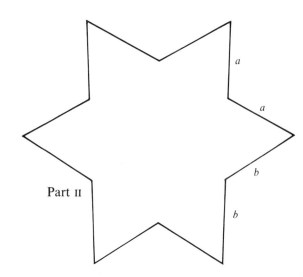

Part II

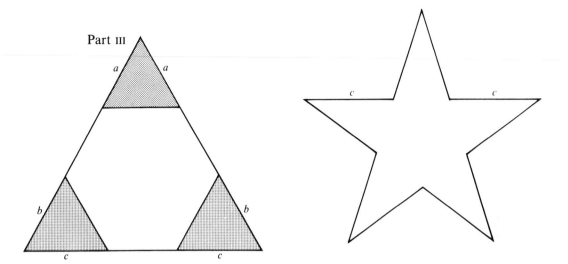

111 Snub dodecadodecahedron

This polyhedron has twelve pentagrams on facial planes that lie above and parallel to twelve pentagons, very much like **73**, only here the pentagrams share edges with sixty equilateral triangles that give it the snub quality. As you can see from the drawings of the facial planes the triangles and pentagons intersect in such a way that tiny slivers appear on one side of the triangles and on all five sides of the pentagons. You will therefore need a great deal of patience to make this model properly.

Begin the same way as in **110**, surrounding W pentagrams with the appropriate scalene triangular dihedral grooves between the star arms, following the same icosahedral arrangement of colours. This is part I of your work. Part II consists of a rather complex assembly of parts that eventually turns out in the shape of a skew-sided equilateral triangle having three grooves running from its vertices toward a point near the incentre and three slivers radiating from near the incentre out toward the sides. The figure opposite will give you some idea of its appearance.

$$|2\ \tfrac{5}{2}\ 5$$
$$12\{\tfrac{5}{2}\} + 60\{3\} + 12\{5\}$$
$$2{\cdot}54887\ 97641$$

Edge length = 11·3 cm
$a = 4{\cdot}0$
$c = 4{\cdot}3$
$d = 3{\cdot}3$
$f = 3{\cdot}7$
$x = 0{\cdot}3$

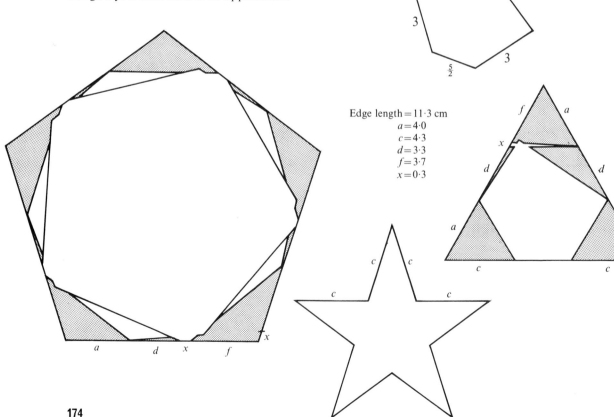

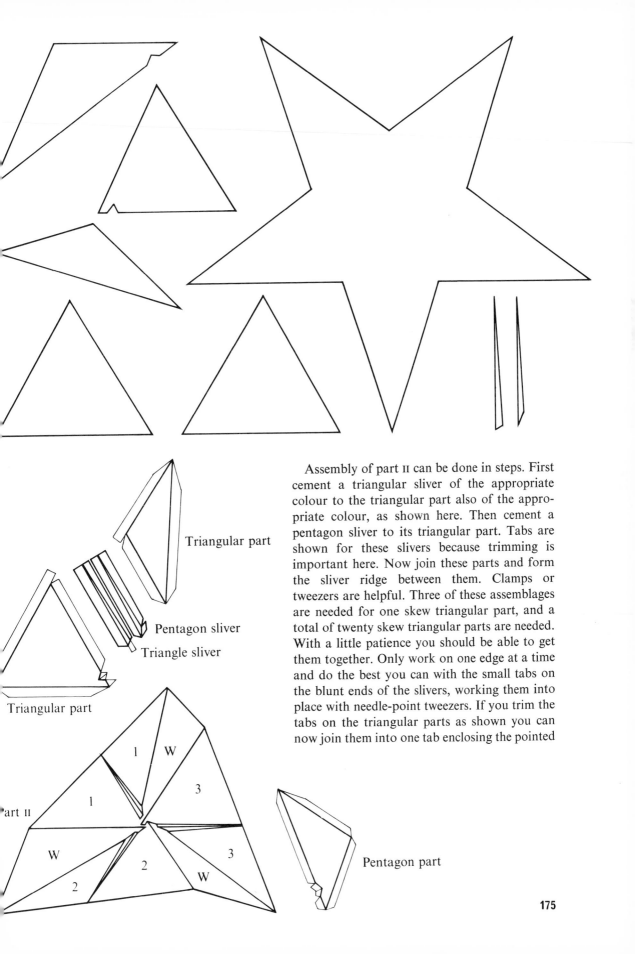

Triangular part

Pentagon sliver

Triangle sliver

Triangular part

Part II

Assembly of part II can be done in steps. First cement a triangular sliver of the appropriate colour to the triangular part also of the appropriate colour, as shown here. Then cement a pentagon sliver to its triangular part. Tabs are shown for these slivers because trimming is important here. Now join these parts and form the sliver ridge between them. Clamps or tweezers are helpful. Three of these assemblages are needed for one skew triangular part, and a total of twenty skew triangular parts are needed. With a little patience you should be able to get them together. Only work on one edge at a time and do the best you can with the small tabs on the blunt ends of the slivers, working them into place with needle-point tweezers. If you trim the tabs on the triangular parts as shown you can now join them into one tab enclosing the pointed

Pentagon part

end of the sliver. This gives added strength and rigidity to this edge. Once you have completed five of part II, they may be cemented in a ring around the (0) section of part I. A colour table follows for all twenty skew triangular parts. The subscript colour is that of the colour sliver. Since pentagon parts are all W they are not in the table.

1	2	3	1	2	3
Y_G	G_O	O_Y	O_R	R_B	B_O
B_Y	Y_R	R_B	R_G	G_O	O_R
O_B	B_G	G_O	G_Y	Y_R	R_G
R_O	O_Y	Y_R	Y_B	B_G	G_Y
G_R	R_B	B_G	B_O	O_Y	Y_B
O_G	G_Y	Y_O	B_R	R_O	O_B
R_Y	Y_B	B_R	O_G	G_R	R_O
G_B	B_O	O_G	R_Y	Y_G	G_R
Y_O	O_R	R_Y	G_B	B_Y	Y_G
B_R	R_G	G_B	Y_O	O_B	B_Y

A simpler approximate model of this polyhedron can be made with the twenty skew triangular parts omitting the slivers. The figure should give you the idea. These can be used as part II and cemented to part I in the way explained above. Approximate facial planes are shown below. The scale of this approximate model may be much smaller, in fact on a small scale it gives good results. See the photograph.

Edge length = 6·7 cm
a = 2·1
b = 2·4
c = 2·5

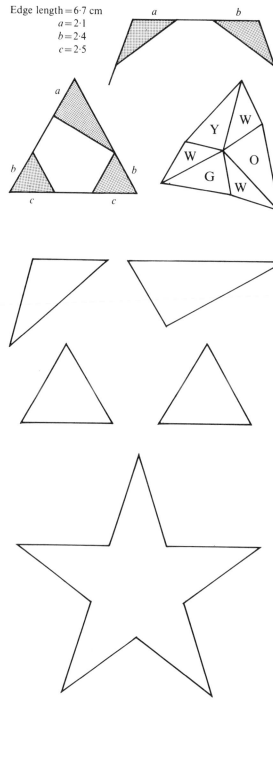

176

112 Snub icosidodeca-dodecahedron

This polyhedron, like **111**, has pentagrams on planes above and parallel to pentagons, but here the pentagrams are twisted with relation to the pentagons, the twist making room for twenty triangles in addition to the sixty triangles sharing edges with the pentagrams.

An alternative method of assembly is best used in making a model of this polyhedron, by cementing the pentagrams last because of the intricate understructure. The work may therefore be done as follows. Use the icosahedral arrangement of colours for the sixty triangles associated with the pentagrams, all of which are W. The other twenty triangles had best be a seventh colour, say T (for tan), because a small central portion of these triangles appears in the bottom of a trihedral dimple whose faces are the corner portions of three different pentagon planes all of which are W. Single polyhedral vertex parts may be constructed in the arrangement shown in the figure. The colour arrangement for the (0) section alone is set out on p. 178. The other sections are in the usual cyclic permutation of colours.

$$|3 \tfrac{5}{3} 5$$
$$12\{\tfrac{5}{2}\} + 60\{3\} + 12\{5\}$$
$$2\cdot25379\ 58256$$

Edge length $= 10$ cm

$a = 3\cdot2$
$b = 3\cdot2$
$c = 3\cdot8$
$d = 2\cdot6$
$e = 1\cdot3$
$f = 1\cdot9$
$g = 2\cdot3$
$h = 1\cdot9$
$k = 2\cdot6$
$p = 1\cdot0$
$q = 1\cdot7$
$r = 1\cdot2$
$x = 1\cdot5$
$y = 2\cdot1$

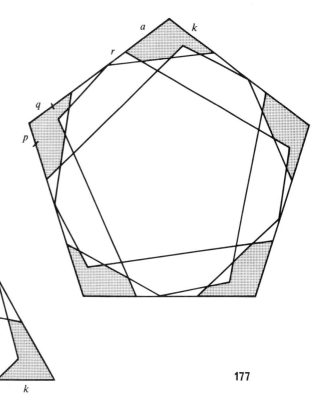

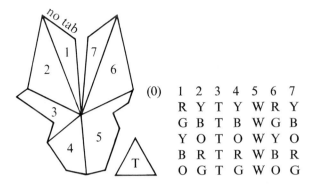

	1	2	3	4	5	6	7
(0)	R	Y	T	Y	W	R	Y
	G	B	T	B	W	G	B
	Y	O	T	O	W	Y	O
	B	R	T	R	W	B	R
	O	G	T	G	W	O	G

Triangles 1 and 7 must be turned up and over 2 and 6, and then the tabs between 1 and 2 and between 7 and 6 are used again as double thickness tabs and cemented to form a vertex part, the centre point being raised in doing this. Triangles 1 and 7 will later become the under surface of one star arm. Five of these vertex parts are joined in a ring to form one section. The tab at the blunt end of triangle 7 is cemented to the upper surface of triangle 1 whose tab is best removed to eliminate the ribs here which would interfere with the pentagram. This pentagram is now added as a cap completing this section.

Clamps can easily be used in doing this since there is an acute overhanging edge to work on. If the pentagrams tend to sag do not be dismayed, gentle pressure can be exerted to straighten them after the cement is well set or after the model is completed. Twelve of these sections complete the model. Admittedly these sections have many jagged edges, but you will be amazed at the way they all fit neatly together.

The alternative method of construction is reminiscent of that used in **83**. If you use this method begin by first cementing the triangles 1 and 7 between the star arms, then fold them under and turn the other long tabs outward to form ribs radiating from the centre out to the star points. Next the triangle pair 2 and 6 form grooves between the solid star arms, the ribs serving as cementing tabs for these triangles. Finally the remaining parts 3, 4, 5 complete the vertex parts around the star points and complete one section as before. This second method of assembly assures a better fit for the pentagrams, because in this method there is less tendency for them to sag.

113 Great inverted snub icosidodecahedron

This polyhedron is another snub that is simpler in construction than most of the others in this set. The reason is that it does not have a crossed vertex figure, so its vertex parts are thereby simplified. Thus a model can easily be made by simply joining the sixty vertex parts following the arrangement set out below, showing one vertex part. Make five such parts and then join them in a ring to form one section. Twelve of these sections complete the model. The central dimples in these sections will remind you of the analogous dimples in the compound of five tetrahedra. Dimples like these appear also in **115** and **116**. The colour arrangement may again be icosahedral. The colours for only one section, the (0) section, are set out below, because the rest follow the usual permutation pattern. All the pentagrams are W.

These sections have very jagged edges and several very small tabs, so you will need patience and care in building this model.

$$|\ 2\ 3\ \tfrac{5}{2}$$
$$(20+60)\{3\}+12\{\tfrac{5}{2}\}$$
$$1\cdot63216\ \ 13496$$

edge length $= 14$ cm
$a = 2\cdot4$
$b = 3\cdot8$
$c = 0\cdot4$
$d = 2\cdot4$
$f = 3\cdot5$
$g = 4\cdot6$
$h = 0\cdot7$
$j = 1\cdot6$
$k = 2\cdot2$
$l = 0\cdot9$
$m = 1\cdot4$
$n = 2\cdot0$
$x = 3\cdot9$
$y = 0\cdot3$
$z = 2\cdot6$
$a+c+h=f$
$k+l+m=g$

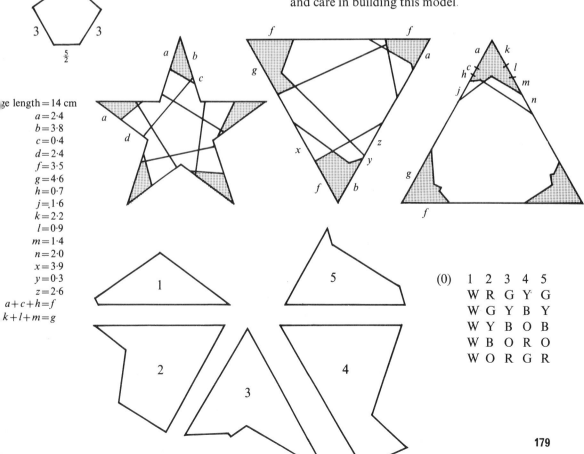

(0)	1	2	3	4	5
	W	R	G	Y	G
	W	G	Y	B	Y
	W	Y	B	O	B
	W	B	O	R	O
	W	O	R	G	R

179

114 Inverted snub dodecadodecahedron

This polyhedron has a very unusual feature in that the star arms of the pentagrams are slightly nicked by the facial planes of the pentagons and one set of triangles. This introduces a very complex structure near the polyhedral vertices. The pentagons are also on planes parallel to the pentagrams but very deep down near the equatorial planes of the polyhedron. The understructure of the pentagrams is reminiscent of that found in **83** and in the snub **112**. This fact suggests a method of constructing a model of this polyhedron, like the alternative method mentioned for **112**.

Begin by cementing triangles 1 and 2 to the star arms of the pentagram (see opposite), then turn them under and cement the other long tabs turned out as ribs. It is not necessary to cut the nicks into the star arms, although after the cementing is completed on triangles 1 and 2 and the solid star has been constructed, the nick in triangle 2 should be cut and removed without severing the rib tab. This will make it easier for the nick in the pentagon corner portion, part 5, to fit later without further cementing, as will be explained shortly. The usual icosahedral colour arrangement is used, with all the pentagrams W.

Your next task is to prepare the parts needed between the star arms. The arrangement of these parts is shown opposite. Here too the usual icosahedral colour arrangement is used, and so only the (0) section is given. The usual permutations apply to the rest.

Tabs are shown because trimming is important.

$$|2 \tfrac{5}{3} 5$$
$$12\{\tfrac{5}{2}\} + 60\{3\} + 12\{5\}$$
$$1\cdot70326\ 04562$$

Edge length = 14 cm

    a = 2·0
    b = 0·7
    c = 1·4
    d = 0·7
    g = 5·3
    h = 3·6
    k = 3·8
    l = 4·2
    m = 3·8
    p = 2·2
    q = 0·7
    x = 2·8
    y = 0·7
    z = 0·4

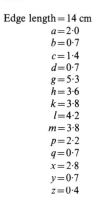

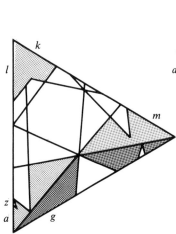

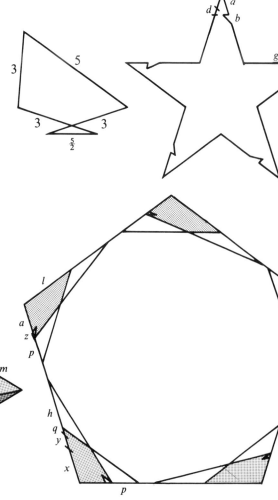

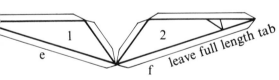

leave full length tab

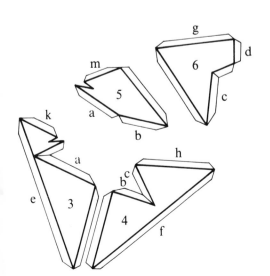

First cement the colour pairs, parts 3 and 4. Then cement part 5 which is W to part 6 which is of an icosahedral colour. Next cement the tabs a, b, c which join these paired parts. A sharp pointed ridge or dihedral angle is thus formed between parts 5 and 6 and the whole assemblage takes on some rigidity. Your final task is to cement these assemblages between the solid star arms.

The ribs under the star take the tabs e and f. Apply the cement to these tabs. Then manœuvre the assemblage into place, and clamp before the cement sets. The small nicked triangle of part 3 dangles at the star point but being attached by

(0)	1	2	3	4	5	6
	Y	Y	Y	O	W	Y
	B	B	B	R	W	B
	O	O	O	G	W	O
	R	R	R	Y	W	R
	G	G	G	B	W	G

181

one continuous tab to the larger portion it is easy to handle. It will get more attention later, but now you must give special attention to the nick in the pentagon corner portion, part 5. Gently ease the upper point over the star arm while the lower point stays below. You need not worry about cementing these since the little nicks have no tabs. The upper point will get further attention later. Once you have placed five of these assemblages between the star arms one section of the work is completed. Twelve sections are needed to complete the model.

Now it is rather a complex task to get the sections joined. The secret is always to cement first the tab at g from part 6 of one section to the tab at h from part 4 of another section. The tabs at d are joined next. Only when the entire model has been completed should you attempt to get the final tabs at k and m adjusted. It is useless to attempt it earlier. You will need tweezers and a probing needle to manœuvre the parts gently into place, then add a drop of cement and fix them with a clamp. The edge here is just acute enough to take a clamp until the cement is set. A good-quality paper 'heals' itself after the clamps are removed or you can help by smoothing it with the probing needle. This model calls for much patience, so good luck to you on this one!

115 Great snub dodecicosidodecahedron

This polyhedron has the special feature of paired pentagrams on common facial planes, found also in **119**. Here the twisted arrangement in the twelve dimples is found, as it is in **113** and again in **116**. However the crossed lines of the vertex figure in this case give this polyhedron a more complex structure for the vertex parts than that found in **113**. The crossed lines appear again in **116**.

To make a model of this polyhedron it will be best to depart from the dodecahedral sections used before and to use an icosahedral assembly technique instead. In this way the crossed facial planes under one star arm can be more easily executed. As for colour the crossed triangular planes (parts 1–5) can well be all one colour, say T (for tan), reserving W for the paired pentagrams and the usual Y, B, O, R, G for the pentahedral dimples, the usual icosahedral arrangement and permutation of colours applying. Notice that in the parts used for assembly purposes triangles 1 and 2 are interchanged from the positions they occupy on the facial planes in the polyhedron itself. This is because these parts will be folded back to form the understructure of one star arm. Join three of the parts 1, 3 to form a

$$|3 \ \tfrac{5}{3} \ \tfrac{5}{2}$$
$$(20+60)\{3\}+(12+12)\{\tfrac{5}{2}\}$$
$$\sqrt{2}$$

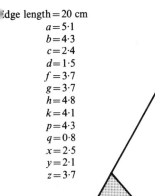

Edge length = 20 cm
$a = 5 \cdot 1$
$b = 4 \cdot 3$
$c = 2 \cdot 4$
$d = 1 \cdot 5$
$f = 3 \cdot 7$
$g = 3 \cdot 7$
$h = 4 \cdot 8$
$k = 4 \cdot 1$
$p = 4 \cdot 3$
$q = 0 \cdot 8$
$x = 2 \cdot 5$
$y = 2 \cdot 1$
$z = 3 \cdot 7$

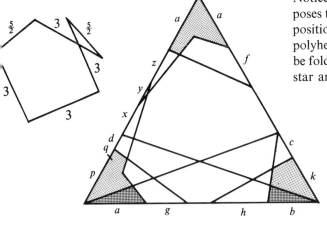

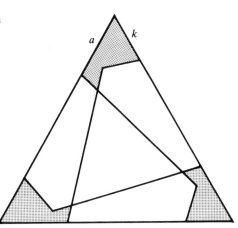

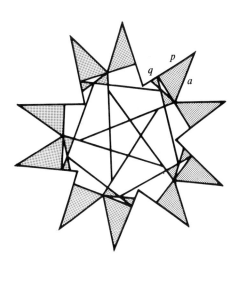

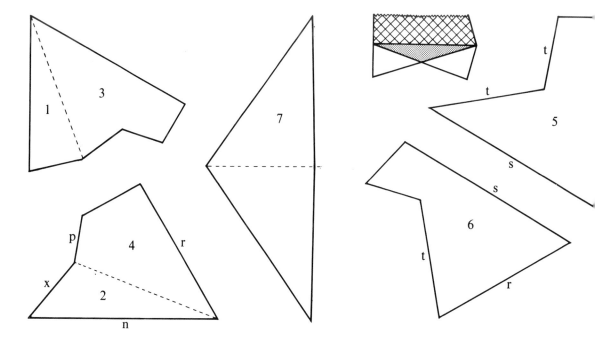

sort of three-bladed propeller. Leave tabs all around, except at x. This tab may be removed because no cementing is done along this edge. Triangle 1 folds neatly under a star arm, as explained later.

Next prepare three pairs of the pentagram parts shown as part 7. Fold the lower star arm shown in fig. (*a*) up and then cement the cross-hatched part shown in fig. (*b*) to the under-surface of the upper star arm allowing the lower part to protrude as shown in fig. (*c*).

The shaded area in fig. (*b*) and (*c*) will later be hidden, so it need not be marked in any way. In fact it is best not to score the paper for this area but only for the tabs. Although the tab at h is cut it will serve as a single tab, as will now be explained.

Three of these pentagram parts must now be cemented around the three-bladed propeller. First join the tabs h, allowing the lower star arm of fig. (*a*) to dangle forward, that is, above the propeller blades. As soon as the cement has set along these edges give the pentagram parts a sharp crease downward. You will then be able to bring the tabs k into contact, so they may be cemented. Now the tab m of triangle 1 on the propeller blades is ready to be cemented to the

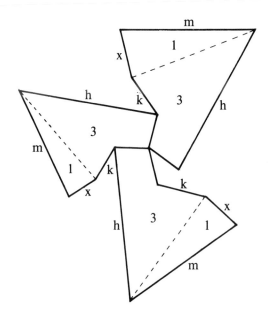

dangling star arm. The sharp overhanging edge along the star arm makes clamping possible while the cement is setting. When you have done this on all three star arms you should have a very rigid assembly, the central portion of an icosa-hedral section which already shows the vertex points of three adjacent polyhedral vertices.

The next step is to cement three of parts 2, 4 to the appropriate edges of the icosahedral assemblage you have just completed. Join the tabs n. Once the cement is set, triangle 2 is folded under the star arm, to bring the tabs p into contact so that they can be cemented. You will now see the intersection of the crossed triangular planes clearly formed where the fold between parts 1, 3 and between 2, 4 occur. Tabs k and p hold these planes in rigid position so no further joining at the folds is needed. No tab is required at x of triangle 2, since the same situation occurs here as at x of triangle 1. You will now also see how the shaded area in fig. (b) is hidden.

The last step of the work is now simple to execute. Add the parts 5 and 6 to complete each of the three vertex parts of the section, matching the tabs q, r and s. Part 5 is T in colour, like the parts 1, 3, since it belongs to the same facial planes, but part 6 in this first assemblage should be Y. In fact all three of part 6, completing each of the three vertex parts of this first assemblage, are Y. Since a total of twenty such assemblages are needed to complete the model, the colours Y, B, O, R, G serve in turn for part 6, three of the same colour to each assemblage.

You will find it best to cement the sections to each other as you complete them, first a ring of five with the (0) arrangement of colours in the dimple in the centre of the ring. Always cement the longer tabs first, beginning with tab t of part 5 from one assemblage and cementing it to tab t of part 6 from an adjacent assemblage. The other tabs then fall readily into place. In doing the last, the twentieth section, leave the tabs s without cement, so these parts on the sixtieth vertex part can be temporarily folded back giving you room to work with a probing needle or tweezers on the other tabs. Cement the tabs s last of all.

This should be a very successful model. The method of construction suggested here can lead to remarkably good results. If you work alone it will take you about 40 or 50 hours to complete the task!

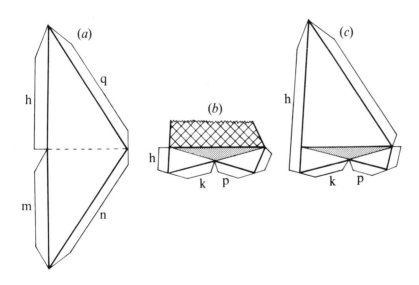

116 Great snub icosidodecahedron

This polyhedron has some very small areas which must technically be considered as the exterior portions of its facial planes, but they are so small that the drawings shown below do not reveal them. Some features of **115** repeat themselves here, the twelve pentahedral dimples, the sixty polyhedral vertices, but instead of paired pentagrams single pentagrams intersect each other in a similar icosahedral fashion. This suggests the use of the same assembly technique in making a model of this polyhedron as that used in **115**. A few compromises will be introduced in the construction work for the sake of practicality, compromises which barely betray themselves in the finished model.

Some enlargements are set out opposite to reveal the detailed parts of the facial planes and the parts numbered for reference.

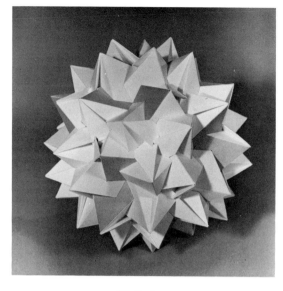

$$|2\ 3\ \tfrac{5}{3}$$
$$(20+60)\{3\}+12\{\tfrac{5}{2}\}$$
$$1{\cdot}29004\quad 04746$$

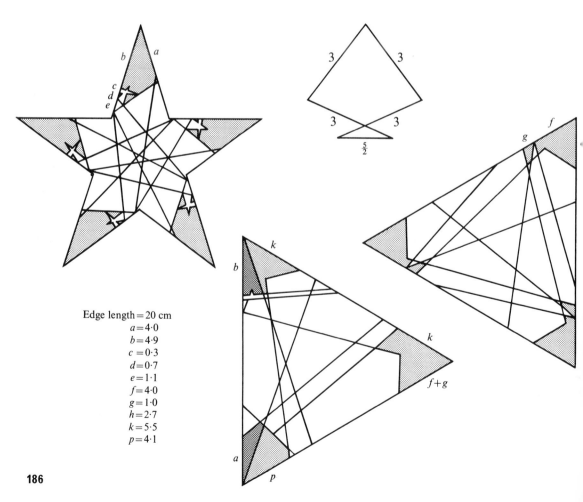

Edge length $= 20$ cm
$a = 4{\cdot}0$
$b = 4{\cdot}9$
$c = 0{\cdot}3$
$d = 0{\cdot}7$
$e = 1{\cdot}1$
$f = 4{\cdot}0$
$g = 1{\cdot}0$
$h = 2{\cdot}7$
$k = 5{\cdot}5$
$p = 4{\cdot}1$

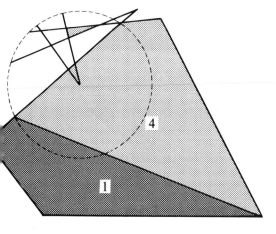

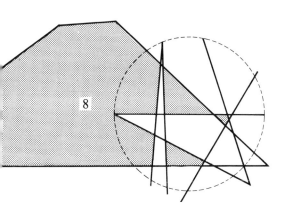

The parts for constructing this polyhedron are set out below, each part identified by a number and each tab by a letter. These parts form an icosahedral section or assembly, like that in **115**, but this one is slightly more complex. The colour arrangement for one section follows; the rest are derived by the usual permutations of colours. The use of the same colour for parts 3 and 5 cannot be avoided since these belong to the snub triangle faces.

1	2	3	4	5	6	7	8
Y	B	B	Y	B	T	T	W
Y	R	R	Y	R	T	T	W
Y	G	G	Y	G	T	T	W

Leave tabs around all parts except at edges marked x, where no tabs are needed. Begin by cementing the pairs 1 and 3, and 4 and 2. Then cement part 7 to part 1, at edges marked p. Part 8 is ingeniously contrived to get a small spiked wedge to raise its ridge above the star arm or pentagram plane. Both sides of the wedge are small triangles, one of which actually belongs to the snub triangle plane and technically should be one of the five colours. Technically it also cuts slightly into part 4. However, the compromise of making both triangles W and letting the wedge touch part 4 without cutting it simplifies the work of construction, and you will see later that this cannot readily be seen because it is

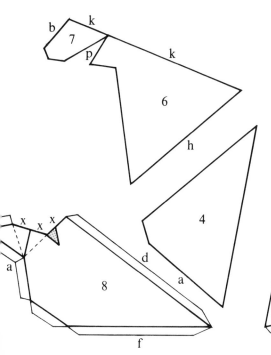

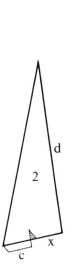

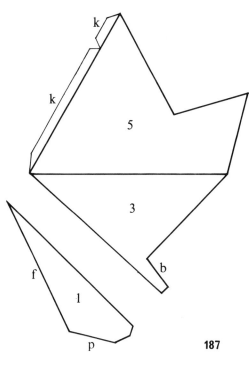

well hidden under a star arm in the completed model. Also, although tab a of part 8 is cut to shape the wedge its two segments are treated as one tab and cemented to part 4 as shown. The next step is to fold up parts 1 and 7 and to cement tab b of part 3 to tab b of part 7. This forms a small but deep trihedral cup, the bottom of which should technically be closed with a tiny triangle coming from the snub triangle plane and a tiny quadrilateral from the pentagram plane. However, omitting these again simplifies the work. So simply cement the tiny tabs at the bottom of part 1, 3, and 7 as best you can and let the trihedral cup have a roughly pointed bottom. This will never be seen in the completed model.

Parts 2 and 8 are now similarly folded up and tab c of part 2 is cemented to tab c of part 8. The two very tiny shaded areas, one on part 2 and the other on part 8 need not be cut or marked. You will see after folding that the wedge butts against the shaded area of part 2 and the edge at x of part 2 crosses into the shaded area of part 8 but no cementing is needed at these places. This folding also forms a trihedral cup, this one both technically and actually pointed at the bottom or base of part 8. The cup formation gives both these partially completed sections a fair amount of rigidity.

The next step calls for some very intricate work, almost impossible to describe in words without an illustration but not impossible to execute. You will have to determine for yourself, by trial and error, the best way to manipulate the parts mentioned in the following description. Take the B, Y, W section of parts 2, 4, 8 and join the R, Y, W section of the same parts 2, 4, 8 to it, matching the tabs d of parts 2 and 8. Let the cement set and give the edge between the B and W a good crease downwards. Then join the third, the G, Y, W section to the R, Y, W section in the same way, tab d to tab d. Let the cement set and give the edge between the R and W a good crease downwards. Now manœuvre the final G and W edges into position and cement the tabs d, using clamps if necessary at the sharp overhanging edge. If you have been

successful you should now see the central portion of this first icosahedral section well formed. Three vertex points become evident, which will eventually belong to three adjacent polyhedral vertices, and a sort of three-bladed pin wheel formed in the centre of the section by the three intersecting edges of three star arms. You should now also see the ridges of the little wedges at the base of the star arms fairly well in line with the edges just cemented.

The next step is much simpler. Cement together tabs f of parts 1 and part 8, making sure that the colour of part 3 corresponds to that of part 2, B with B, R with R, G with G, around the pin wheel section. As soon as the cement is set give these edges a good crease down. This should bring the tabs, already cemented, between parts 1 and 3 into contact with the tabs between parts 2 and 4 behind or below the star arms. These tabs now serve as double-thickness tabs and are cemented to form the line of intersection of the snub triangle planes which cross behind the star arm.

The final step is the simplest of all. Cement parts 4 and 6 at the tabs marked h and join parts 5 and 6 at the tabs marked k. The tab k of part 5 is broken into two segments; the smaller tab will be cemented later on to its mate on part 7 when the sections are joined. When this has been done the three vertex parts of one section are complete. You may find that the central part of the pin wheel is not entirely rigid. This is due to the fact that no cementing was done here. As the model develops and the sections are added one after the other, they exert their own pressure inwards and the final result is satisfactory. In joining the sections, always cement the longer tabs first. Five sections form a ring with the pentahedral dimple in the (0) arrangement of colours in the centre. The remaining tabs require some care and patience, some being so small that you may find they do not need cement. The neatness of the finished model depends on the attention you give to each detail. A total of twenty sections are needed to complete the model. If you do all the work alone you may expect to spend about 50 hours on this one!

117 Great inverted retrosnub icosidodecahedron

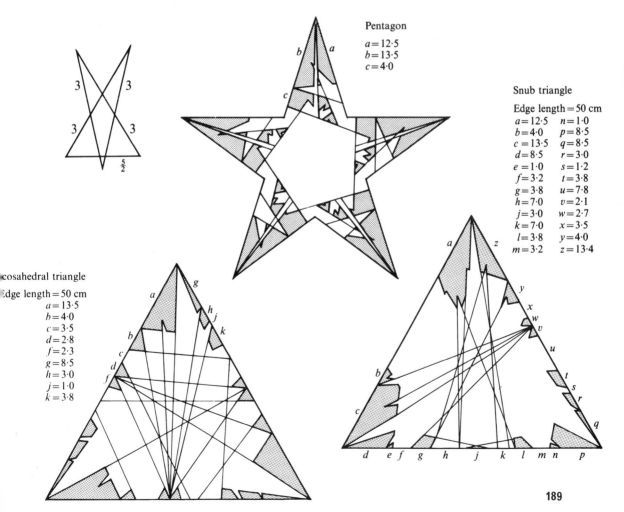

This polyhedron is truly remarkable in its complexity. Deep pentahedral cups display deeply recessed decagrammic rosettes which close off their tapered central portions. These cups have steep outside facial planes that are fantastically intricate. The vertex figure shows how two triangle faces meet just beyond the central portion of a star arm. This introduces very slender spiked wedges whose vertices coincide with the vertices of the pentagrams and which then continue down toward the central portions of the pentagram faces but are cut into two more segments before disappearing into the interior of the solid. You will see this better from the model once you start making it than from any description of it in words. So if you want to attempt this one, here is an assembly technique.

$$|2\ \tfrac{3}{2}\ \tfrac{5}{3}$$
$$(20+60)\{3\}+12\{\tfrac{5}{2}\}$$
$$1\cdot16000\ 30093$$

Pentagon

$a = 12\cdot5$
$b = 13\cdot5$
$c = 4\cdot0$

Snub triangle

Edge length = 50 cm

$a = 12\cdot5$	$n = 1\cdot0$
$b = 4\cdot0$	$p = 8\cdot5$
$c = 13\cdot5$	$q = 8\cdot5$
$d = 8\cdot5$	$r = 3\cdot0$
$e = 1\cdot0$	$s = 1\cdot2$
$f = 3\cdot2$	$t = 3\cdot8$
$g = 3\cdot8$	$u = 7\cdot8$
$h = 7\cdot0$	$v = 2\cdot1$
$j = 3\cdot0$	$w = 2\cdot7$
$k = 7\cdot0$	$x = 3\cdot5$
$l = 3\cdot8$	$y = 4\cdot0$
$m = 3\cdot2$	$z = 13\cdot4$

Icosahedral triangle

Edge length = 50 cm

$a = 13\cdot5$
$b = 4\cdot0$
$c = 3\cdot5$
$d = 2\cdot8$
$f = 2\cdot3$
$g = 8\cdot5$
$h = 3\cdot0$
$j = 1\cdot0$
$k = 3\cdot8$

189

Begin work with the interior parts of one pentahedral cup. Throughout the following description it is assumed that you will be able to recognize the parts by their shape and to see where they are found in their respective facial planes. As for colour, all pentagram planes are W, one set of triangles are all T, and the other set of triangles take the icosahedral arrangement of five colours. Only one section of part I is shown. First prepare the rosette parts and cement them to their respective larger parts at tabs marked z. Fold the rosette parts alternately up and down, up between the central Y and T parts whose tabs a and b then fit the corresponding tabs on the larger Y part. Tab c then matches tab c. Prepare five of these sections in the (0) arrangement of the colours, the rosettes following their own cyclic permutation. The colour O is shown, so repeating this with the others the order is: O R G Y B, while the larger parts follow the order: Y B O R G respectively. The five sections are cemented in a ring, the tab at x from one is cemented to the tab at y from the neighbouring part, and so on around. This completes part I: the pentahedral cup with the decagrammic rosette at the bottom.

One of the five sections of part II is laid out opposite. The W parts belong to one star arm, the T and B parts between them form a slender spiked wedge cutting through the face of the star arm. First cement these four upper parts. Notice that the W part on the left has a shaded area joining a small attached triangle. This should not be cut. Later the second segment of the spiked wedge will be cemented at x to cover this area. But before doing this it is easier to cement the paired triangles of colour B into place along the tabs marked a, b, d, f. Next the second segment of the spiked wedge is cemented in place. When this is done you will see how this spiked wedge seems to penetrate the paired triangles through their shaded areas which should also be left uncut. The paired triangles meet at a very acute angle along their common edge and they are held in a rigid position by the V-cut near the base of the star arm. Next cement the Y triangle at tabs marked l and m. When this is done turn the

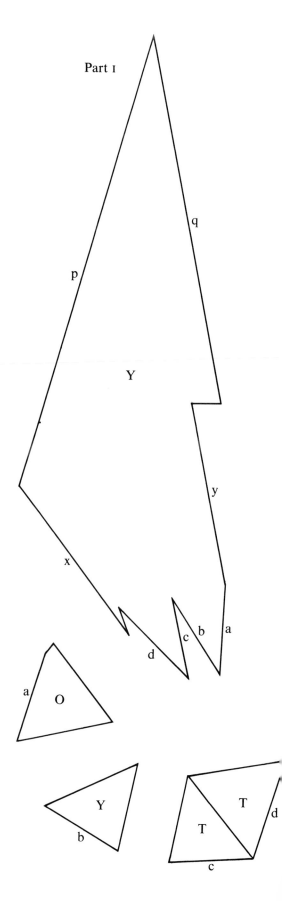

Part I

Part II

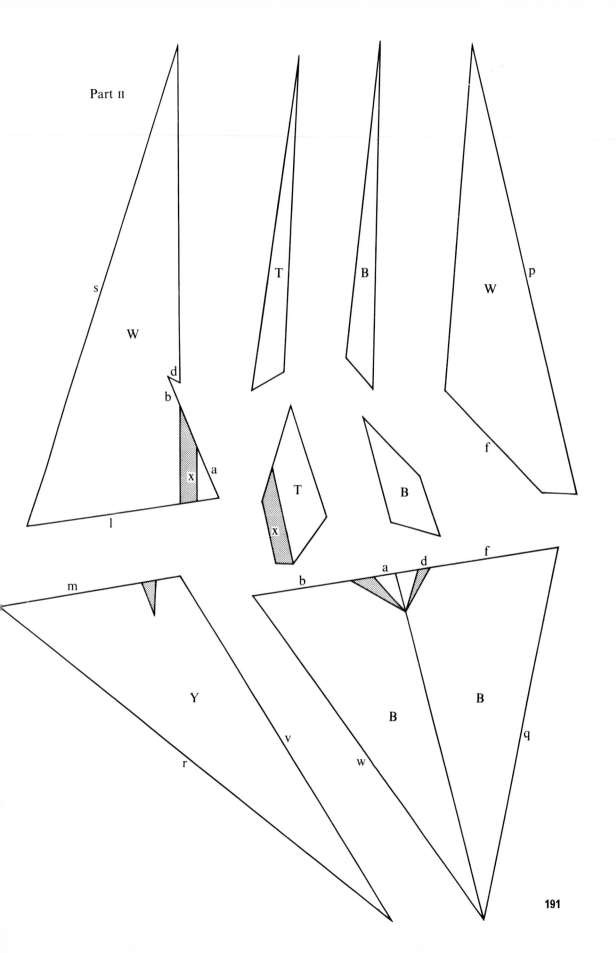

191

Part III

192

Y triangle up and cement the next set of tabs, v and w, of Y to B. You will then see how the end part of the second segment of the spiked wedge covers the shaded area of the Y triangle, so this should also be left uncut. Repeat this section five times, each in its permutation of colours. The five sections are then cemented to part I, first by the tabs p and when these are set, by the tabs q. You will see that part II has the effect of filling the spaces between the steeply projecting points of part I, revealing more clearly but still in an unfinished way a ring of five polyhedral vertices. These polyhedral vertices will be finished or completed by the addition of part III, now to be described.

One of the five sections of part III is laid out opposite. All the shaded areas should be left un-cut because various wedges will cover and so hide these areas. Also full tabs are shown for the W part because these will be needed for joining sections and it is important to have them done correctly. Notice that the largest part, B in colour, has a smaller B part attached to it. This means it can be cut as one piece of paper, but a T part comes between the two small B parts, as shown. The W part will give some rigidity to the T and B upper parts, shaping them into a shallow groove. Once you have cemented these parts, the two lower wedges are assembled and cemented as shown, the shaded areas marked y are cemented to each other and then the shaded areas marked z. These wedges can be done as small irregular polyhedra in their own right, but generally their end faces need not be closed. The T and Y wedge is the third segment of the slender spiked wedge lying on the face of the star arm, the paired O parts will eventually be the continuation of the paired O triangles, like the paired B triangles in part II. This section is repeated five times using the appropriate permu-tation of colours. Then each is cemented so tabs s and r of part III go with those at s and r of part II. A pair of very tiny triangles forming a very small wedge is shown in part III, but this need not be added until further sections are completed. In fact it is so small that for practical reasons it could be omitted and never be missed.

This completes one section, namely one ring of five polyhedral vertices. Twelve such sections are needed to complete the model. To join these cement the tab h in part III of one section to tab g of another section. A third section added to the first two will show that this portion of the pentagram planes near the tabs h and g forms a shallow trihedral dimple, occurring twenty times on the completed model. Your patience and perseverance will have to hold out for more than 100 hours if you want a complete model of your own.

However, you can introduce some com-promises to simplify and shorten the work. You might for example design the parts to omit the decagrammic rosette and all the wedges. Your model will then have the same vertices as the original, but only some of the major portions of the facial planes will be left. A set of simplified parts is shown below, together with a photograph of this simplified model.

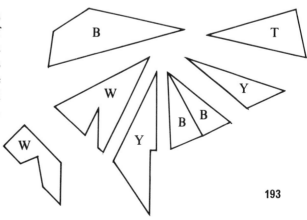

118 **Small inverted retrosnub icosicosidodecahedron**

This complex polyhedron has one thing in common with **110**, namely it has facial symmetry as you can easily see from the drawings of the facial planes. But here you will notice a great deal more complexity. Hence for this model, as for the previous one, **117**, you will need unusual patience and perseverance to complete a model. The assembly method outlined here will make use of only three colours, Y, R, W, one for each of the three kinds of facial planes.

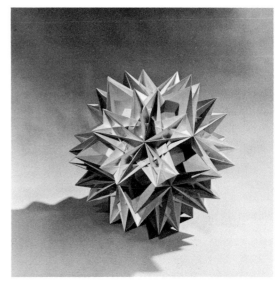

$$\tfrac{3}{2}\ \tfrac{3}{2}\ \tfrac{5}{2}$$
$$(40+60)\{3\}+12\{\tfrac{5}{2}\}$$
$$1\cdot16138\ \ 96003$$

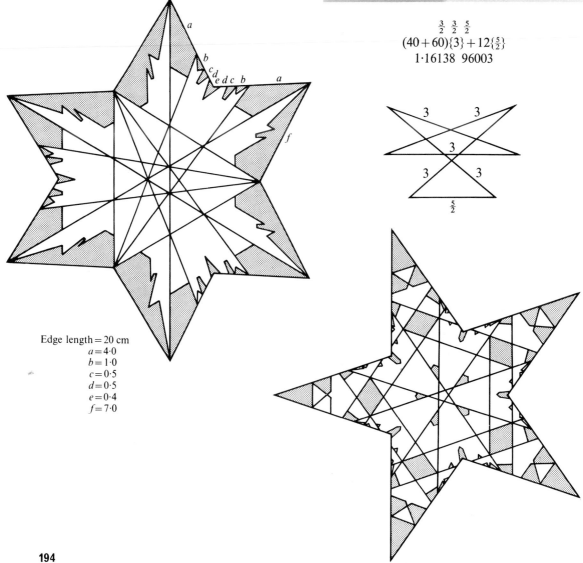

Edge length = 20 cm
$a = 4\cdot0$
$b = 1\cdot0$
$c = 0\cdot5$
$d = 0\cdot5$
$e = 0\cdot4$
$f = 7\cdot0$

194

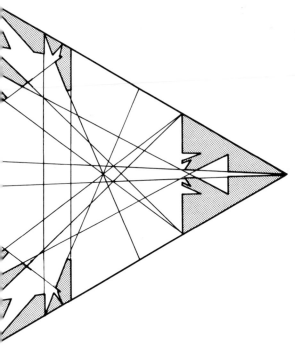

Begin with three sets of the parts shown as part I, cementing the tab a of one part to the tab b of another, then c from one to d from another. After the third set has been added, you will have a deep cup whose trihedral bottom is composed of three W rhombi and its sides of six Y quadrilaterals. Twenty of these will be needed to complete the model.

A good way to proceed with the construction of part II is to begin with the larger parts shown as (a) on p. 196; the layout of pieces is shown on p. 197. The R quadrilateral forms a dihedral groove with the large Y piece. The small R and W pieces form a small wedge and when this is cemented in place at the corresponding tabs g, h, k, it gives some rigidity to the R quadrilateral. The pieces at the bottom of (a) will eventually fold up, but before this is done it is better to assemble all the small pieces in (b). These turn out to be a sort of butterfly embossed on a shallow dihedral groove. The plan for this is shown in (c) where the dihedral groove is cross-hatched. The quadrilateral and the Y triangle at the top of part II (b) can be left uncut as one piece of paper. This Y portion should be given a very sharp fold downwards to make cementing easier or even unnecessary.

Once you have completed the unit shown in

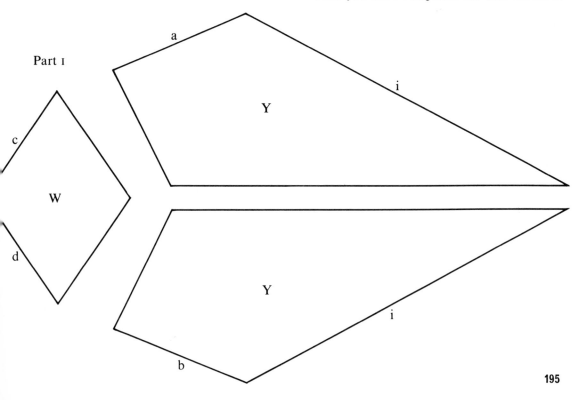

Part I

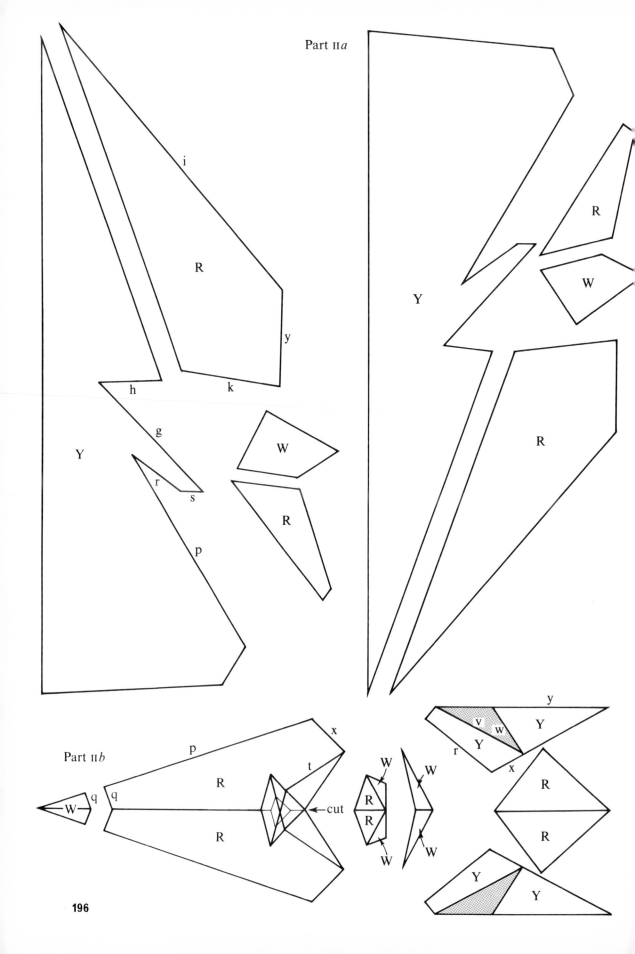

Part IIa

Part IIb

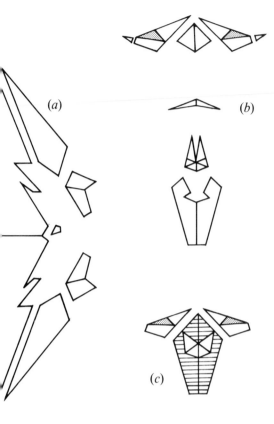

(a)

(b)

(c)

Part II

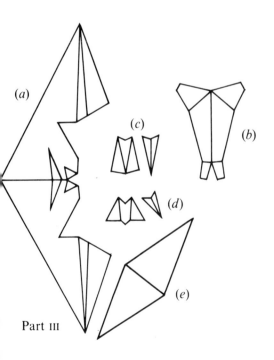

(a)

(c)

(b)

(d)

(e)

Part III

part II (*b*) it is cemented first by tab p of part II (*a*). Turn the Y parts of part II (*b*) sharply down to form a very acute dihedral angle, and cement the tabs at y. You will then find that the tabs v and w can more easily be joined or may not need any cement. The tabs r can be cemented next and then the very small W triangle positioned to close the bottom of the long narrow trihedral hole by matching the tabs marked s. For practical reasons this triangle may be omitted, since it is almost impossible to see from the outside.

The bottom half of part II (*a*) may now be folded up and the appropriate tabs at p are again cemented first, then those at y. Finally the small hole at the bottom is closed with the kite shaped W part, shown in II (*a*), tab q going with tab q at the bottom of II (*b*). All of part II will thus give you a large butterfly section. Three of part II are cemented around each part I joined by the tabs marked i. Since the model calls for a total of twenty of part I, you will need sixty of part II. And part III is still to come. You can of course proceed to part III at once and join parts as they are completed.

One section of part III is set out on p. 198. The pieces of part III (*b*) form a dihedral wedge and this is cemented by tabs a, b, c, d, f to the larger triple R pieces of part III (*a*). It is easiest to begin with tab b, then the R parts with tabs d and f are folded up to form a deep groove surrounding the W part of part III (*b*). The pieces shown in part III (*c*) and (*d*) are wedges. It is best to assemble these as shown, leaving the shaded parts x and y uncut. You can then spread cement over the shaded areas of these wedges and cement them in place over the corresponding areas of part III (*a*). The bottom part of part III (*a*) may now be folded up. The wedges will then be deeply embedded in the narrow groove that is formed.

Repeat the instructions for part III (*a*), (*b*), (*c*), (*d*) to build a second section. Then these two are joined at the tabs marked j, using the piece shown in part III (*e*) as a connector. Part III is now complete. It forms a sort of wedge-shaped sandwich, the 'bread' being two equilateral triangles of colour R, joined at their bases and

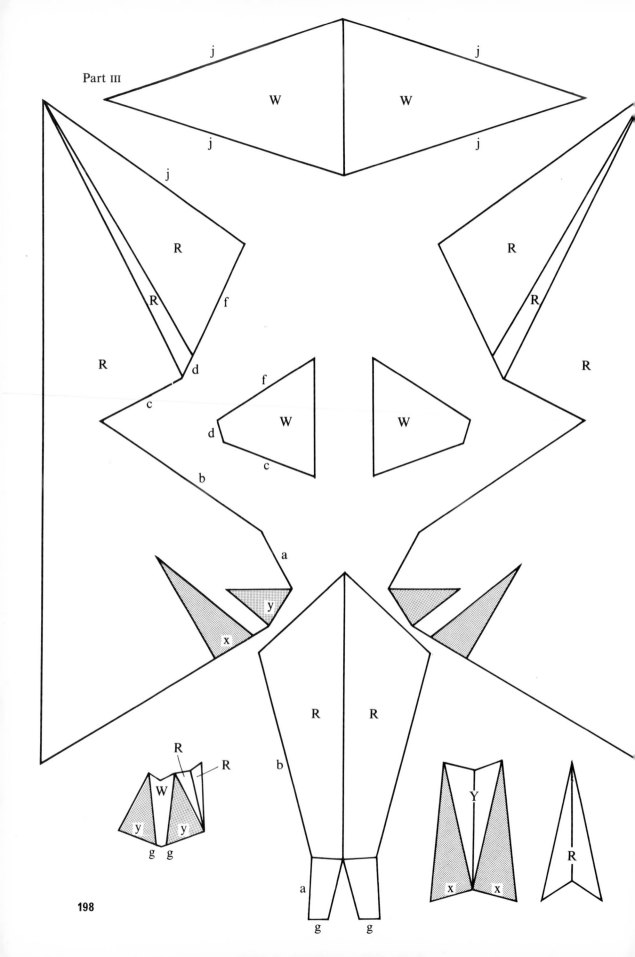

Part III

198

diverging slightly toward their vertices to show the 'meat', an assortment of smaller wedges. You will need a total of thirty of these wedge-shaped sandwiches to complete the model. Once you have completed three of them you can of course cement them around the completed parts I and II. You will then have completed three polyhedral vertices and have the beginnings of three others. As the model grows it is not hard to see how parts I, II, and III are fitted together. Parts II and III alternate in a ring, their edges forming a set of ten lines sloping gently inwards and meeting at the central point of a dodeca-hedral section. This would suggest an alternative assembly technique. The long edges of all the parts make any order of assembly easy.

An alternative colour arrangement is suggested by the twenty cups of part I, namely the usual icosahedral arrangement. The six quadri-laterals of part I could be YB, YG YR, to name colours for only the first cup. The others would be derived from the usual permutations. You could work out the colours for parts II and III by following the facial planes through from part I as the work progresses.

Admittedly this polyhedron will take a long time to assemble. As in the previous model you can expect to spend more than 100 hours on this one. However, some simplifications are possible here also. They are as follows: in part III omit the small wedges of part III (c) and (d), and simply make III (b) pointed at the bottom near tab y of part III (a); in part II omit the embossed butterfly assemblage of part II (c) and make II (c) a simple shallow dihedral groove pointed at the bottom filling in the area near q of II (a). Part I remains the same. This simplified model will have the same vertices as the original and only some small, scarcely noticeable portions of the facial planes will be missing.

119 Great dirhombicosi-dodecahedron

This polyhedron is remarkable in more ways than one. First the symbol $| \frac{3}{2}\ \frac{5}{3}\ 3\ \frac{5}{2}$ shows that it differs from every other uniform polyhedron, in that all the others have a symbol made up of only three numbers, either integers or fractions or both. These three numbers relate them to Schwarz triangles on a spherical surface. The existence of this polyhedron indicates that there is in general no reason for the restriction to triangles. Does this mean that possibly other uniform polyhedra may still be discoverable? A quotation from the paper by Coxeter and others, *Uniform polyhedra* (p. 427), will answer this question: 'We can only say that higher spherical polygons would have to satisfy various conditions which are almost always incompatible.' So the most that can be said is that the list is probably complete with this as the seventy-fifth of the set of uniform polyhedra.

$$| \tfrac{3}{2}\ \tfrac{5}{3}\ 3\ \tfrac{5}{2}$$
$$40\{3\} + 60\{4\} + 24\{\tfrac{5}{2}\}$$
$$\sqrt{2}$$

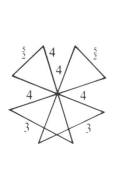

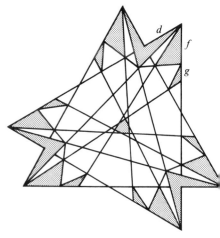

Edge length = 20 cm
$a = 4\cdot2$
$b = 5\cdot2$
$c = 1\cdot6$
$d = 4\cdot1$
$f = 3\cdot5$
$g = 1\cdot8$
$h = 4\cdot2$

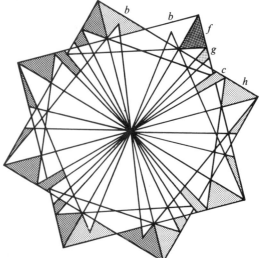

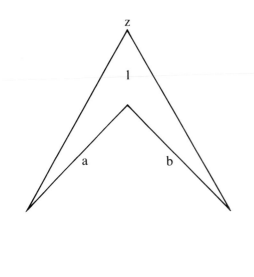

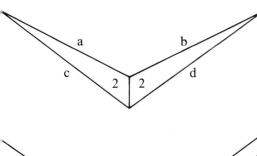

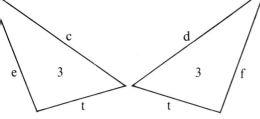

Another way in which this polyhedron differs from all the others is that it is the only known polyhedron that has more than six faces at each vertex. In fact it has eight. These faces occur in coplanar pairs, twelve pairs of pentagrams, twenty pairs of triangles and thirty pairs of diametral squares. The vertex figure shows how they are ordered in rotation at each vertex; the squares occurring alternately with the other faces.

In making a model of this polyhedron a method of assembly by dodecahedral sections is suggested. The central dimples of these sections are reminiscent of similar dimples that occur in the compound of ten tetrahedra, one of the icosahedral stellations. Begin by cementing a ring of five of part 1 in the icosahedral (0) arrangement of colours. You should have no difficulty in identifying this part and the others referred to, from the drawings of the full facial planes. When the ring is completed you will have a dimpled five-pointed star.

It will be simplest to make all the squares one colour, T. So the next step is to cement the small V-shaped pieces, part 2, between the dimpled star arms, first by the tab a and when this is well set, by the tab b. You should now have a lip all around the dimpled star, like the lip on a slip-cap cover. Part 3 is joined next, by the tabs at c and d. The colours of part 3 should correspond to those of part 1. They are easy to get right because you can see them as the continuation of the facial planes of part 1, half a dimpled star arm forming a dihedral wedge apparently protruding through these planes. When this is done cement tabs t and you will see that the spaces between the dimpled star arms have been filled; the outer edges at tabs e and f form a skew decagon. This completes the dimpled star, the central portion of one dodecahedral section.

Continue now with the assembly of parts 4, 5, 6, 7 and 8. Parts 4, 5, 7 all belong to the facial planes of the squares, so their colour is T. Dotted lines are scored on the reverse side and the fold is made upwards. Part 6 belongs to the triangle planes and hence it will require colour

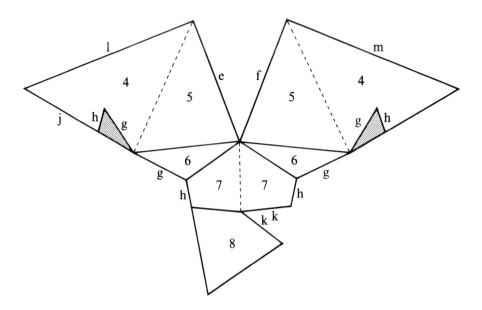

pairs. Although the drawing shows the 6 joined to 5, and 7 to 8, these parts must actually be prepared separately because they are different colours. The colours for 6 are not hard to get right once you see the permutation pattern. For the first set of five, the (0) set, they go like this: YR BG OY RB GO. The order is Y, B, O, R, G for the left member of each pair, starting at the beginning of the line. Then, Y, B, O, R, G for the right member but starting at the third pair and continuing the cyclic order from the end of the line to the beginning. The other sections later follow the same pattern. For example the (1) section is: BO YG RB OY GR where the cyclic order is B, Y, R, O, G. And so on for the rest.

In constructing this section, leave tabs all round for all parts. The shaded area of part 4 need not be scored or marked in any way and it needs no tab. The next step is, as in the case of **116**, difficult to describe in words without actual demonstration, but some trial and error on your part should lead to success. Start with the left-hand side. Give parts 6 and 7 a good crease downwards to form a triangular wedge or di-hedral angle. Spread a drop or two of cement on the tabs at g and h. Then fold 4 and 5 up and manœuvre the tabs g and h on to the shaded area of 4. Clamp with tweezers and allow the cement to set. Then perform the same operation on the right. If you have succeeded you will now have a

sort of butterfly section, deeply grooved between parts 4 and 5, with 6 and 7 forming a wedge apparently penetrating the surface of 4 through the shaded area. Part 8 is now added, by joining the tabs at j and k. Part 8 is W, since it belongs to one of the paired pentagram planes. All these will be the same colour, W.

When you have completed five of these butter-fly sections cement them around the dimpled star, tabs e and f of part 5 to e and f of part 3; namely at the skew decagon edges. Cement one set of tabs at a time. Say those at e first, let them set up well, then give the butterfly section a good crease downwards and cement the tabs at f. You can now use clamps at these edges because there is an acute dihedral angle.

As this is completed you will see the outer edges forming a new skew decagon at the tabs l and m. You will also see that the fold between parts 4 and 5 from one butterfly section come into contact, or nearly so, with the corresponding fold in an adjacent section. At this stage you might want to join these parts along the line of contact, but this is not really necessary. Star arm sections must still be added, as will be explained presently, and then a third line of contact will appear in the same place. Since all this will eventually be hidden inside the model and since the completed model will have suffi-cient rigidity without joining these parts along this line of contact, it simplifies the work to proceed to the next step.

The star arm sections, parts 9 and 10, belong to the paired pentagram planes and part 11 to the paired squares. So 9 and 10 are both W and 11 is one piece, colour T, scored along the dotted line. Leave tabs all around as usual. The tabs at n and p are of special design. The protruding segments will later join one section to the next to form a single tab with its neighbour. The fold in 11 forms the third line of contact referred to above.

This star arm section is very simple. First it will be best to cement the tabs q and r to form a deep dihedral groove. Then cement this section by tabs l and m to tabs l and m of the two parts 4. Cement l first and, when this is set, cement m. Once you have done this all around the skew decagon edges, the protruding tabs at n and p are joined across the pointed end of part 8. This completes one dodecahedral section. The joined tabs, n and p, now give this section distinctly pentagonal edges, except that the corners have V-cuts left from part 9. If you were to complete three dodecahedral sections, joining them along the tabs n and p, you would see that a hexagonal hole would be left centrally between these

sections. This central hole is closed with the section composed of parts 12 and 13. The six quadrilaterals, part 12 (all T) belong to six different paired square planes, forming the sides of a cup which is pointing directly toward the centre of the polyhedron. It is cut off, not very far down, by the isosceles hexagon, part 13 (colours Y, B, O, R, G), which belongs to the exact centre portion of the paired triangle planes.

The best procedure here is to make five of these cups (twenty in all will be needed) and to add them immediately to the (0) dodecahedral section just completed. You can determine the colour of 13 by aiming the point z of part 13 at the point z of part 1, and thus getting the colours to correspond.

Twelve dodecahedral sections are needed for the complete model. Always fill in the V-cuts left by part 9 around these sections as each part is cemented in place. The assembly will remind you of the way in which the regular dodecahedron is assembled, but what a difference! Where the regular dodecahedron has twelve faces, here there are twelve multifaceted sections of a most intricate design. Truly, a remarkable polyhedron! It will take you at least 50 hours work to complete this model.

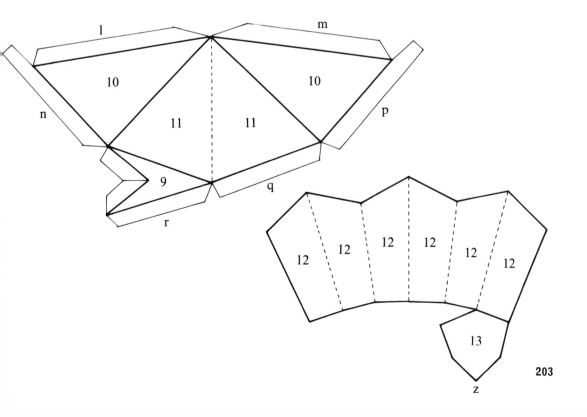

A final comment

Even if you have now made only a few of the non-convex uniform polyhedra, you can see from the models what properties belong to the set as a whole. The most interesting fact is that all of them are derived from Schwarz triangles—except one, model **119**. It is exceptional in another way. It is the only known polyhedron that has more than six polygons surrounding each vertex, four squares alternating with two triangles and two pentagrams. All the squares are on planes through the solid's centre of symmetry. It is classified as a snub polyhedron because here the squares may be regarded as snub faces instead of the usual triangles as in other cases. The existence of this polyhedron indicates that spherical polygons as well as spherical triangles may give rise to other uniform polyhedra. However it is a complex task to investigate the possibilities. It still remains to be done.

You may be wondering why the stellation process was treated so thoroughly in Section II. This was done, first of all, because in some ways it is breaking new ground. Secondly, it is intrinsically a simple process, although it may indeed lead to polyhedral forms almost too numerous to detail. With enough perseverance you can discover any number of these forms by yourself. Lastly, it should help you to understand another kind of stellation, namely edge stellation. Edge stellation is that in which the edges of a polyhedron are produced to generate the edges of a new polyhedron. A simple example is found in the dodecahedron, whose edges if produced generate the edges of the small stellated dodecahedron. Stick models can show this very plainly. Many uniform polyhedra are edge stellations of other uniform polyhedra. But it must be left for you to pursue this matter further on your own.

Epilogue

This book has presented only some polyhedral forms. For anyone acquainted with the field there are obvious omissions, the two infinite sets of prisms and antiprisms, all the Archimedean duals (except for the three given on pp. 6–8), and many other polyhedral forms. Among the Archimedean duals two are especially noteworthy, the rhombic dodecahedron and the rhombic triacontahedron. The former is given in Cundy and Rollett along with the stellated forms worked out by Dorman Luke. Stellated forms of the latter are presented in summary fashion without drawings except for the stellation pattern by J. O. Ede in the *Mathematical Gazette*, XLII (1958). All the Archimedean duals can be stellated, as indeed any polyhedron can. In the light of what you have now learned, you can discover the stellation patterns by yourself, and thus make models of all these polyhedra using the methods and techniques described. So

beyond the models presented here, there are more, and more and more! The object of an investigator would not be to multiply forms but to arrive at the underlying mathematical theory that unifies and systematizes whole sets of polyhedral forms.

From this point of view the mathematical investigation takes its origin from an inductive process, akin to the scientific method; namely, to observe individual instances of any phenomenon, then to classify and systematize in order to arrive at general principles which serve as the basis of a deductive process. Many people are not aware of this aspect of mathematics, but the history of mathematics is full of instances bearing this out. (See G. Polya, *Mathematics and plausible reasoning*.)

So, to end on the same metaphor as that used in the preface, the road still stretches on before you. Why don't you continue your journey?

A faceted form of the small stellated dodecahedron
(from Bruchner: VIII, 14)

A stick model of the icosahedron

A stick model of the dodecahedron

References

Brückner, M. *Vielecke und Vielflache*. Tuebner, 1900.

Coxeter, H. S. M. *Introduction to geometry*. John Wiley and Sons, 1961.

Coxeter, H. S. M. *Regular polytopes*, 2nd ed. Macmillan, 1963.

Coxeter, H. S. M., 'Polyhedra', chapter 5 in Ball, W. W. R. *Mathematical recreations and essays*. Macmillan, 1965.

Coxeter, H. S. M., Du Val, P., Flather, H. T. and Petrie, J. F. *The fifty-nine icosahedra*. University of Toronto, 1951.

Coxeter, H. S. M., Longuet-Higgins, M. S. and Miller, J. C. P. 'Uniform polyhedra'. *Phil. Trans.* 1954, **246**A, 401–50.

Cundy, H. M., and Rollett, A. P., *Mathematical models*, 2nd ed. Oxford, 1961.

Ede, J. D. *Mathematical Gazette* (1958), XLII.

Heath, T. L. *Euclid's elements*, vol. 3. Dover, 1956.

Hilbert, D. and Cohn-Vossen, S. *Geometry and the imagination*. Chelsea, 1956.

Lines, L. *Solid geometry*. Dover, 1965.

Lysternick, L. A. *Convex figures and polyhedra*. Dover, 1963.

Polya, G. *Mathematics and plausible reasoning*.
Vol. I. *Induction and analogy in mathematics*.
Vol. II. *Patterns of plausible inference*. Oxford, 1955.

Steinhaus, H. *Mathematical snapshots*. Oxford, 1960.

Wenninger, M. J. *Mathematical Gazette* (1968), LII.

Wenninger, M. J. *Polyhedron models for the classroom*, N.C.T.M. Publication, 1966.

List of Models

The figures in parentheses refer to the number of the model in the paper by Coxeter *et al.* 'Uniform Polyhedra'.